THE ORGANIC CHEMISTRY OF DRUG SYNTHESIS

VOLUME 3

THE ORGANIC CHEMISTRY OF DRUG SYNTHESIS

VOLUME 3

DANIEL LEDNICER

Analytical Bio-Chemistry Laboratories, Inc.
Columbia, Missouri

LESTER A. MITSCHER

The University of Kansas School of Pharmacy
Department of Medicinal Chemistry
Lawrence, Kansas

A WILEY-INTERSCIENCE PUBLICATION

JOHN WILEY AND SONS

New York · Chichester · Brisbane · Toronto · Singapore

Library of Congress Cataloging in Publication Data:
(Revised for volume 3)

Lednicer, Daniel, 1929–
  The organic chemistry of drug synthesis.

  "A Wiley-Interscience publication."
  Includes bibliographical references and index.
  1. Chemistry, Pharmaceutical. 2. Drugs. 3. Chemistry,
Organic—Synthesis. I. Mitscher, Lester A., joint
author. II. Title. [DNLM 1. Chemistry, Organic.
2. Chemistry, Pharmaceutical. 3. Drugs—Chemical
synthesis. QV 744 L473o 1977]

RS403.L38   615'.19   76-28387
ISBN 0-471-09250-9 (v. 3)

Printed in the United States of America

10 9 8 7 6 5 4 3 2

With great pleasure we dedicate this book, too, to our wives,
Beryle and Betty.

The great tragedy of Science is the slaying of a
beautiful hypothesis by an ugly fact.

Thomas H. Huxley, "Biogenesis and Abiogenisis"

Preface

The first volume in this series represented the launching of a
trial balloon on the part of the authors. In the first place,
we were not entirely convinced that contemporary medicinal
chemistry could in fact be organized coherently on the basis of
organic chemistry. If, however, one granted that this might be
done, we were not at all certain that the exercise would engage
the interest of others. That book's reception seemed to give
an affirmative answer to each of these questions. The second
volume was prepared largely to fill gaps in the coverage and to
bring developments in all fields up to a common date - 1976.
In the process of preparing those volumes, we formed the habit
of scrutenizing the literature for new nonproprietary names as
an indication of new chemical entities in or about to be in the
clinic. It soon became apparent that the decreased number of
drugs being granted regulatory approval was not matched by a
decrease in the number of agents being given new generic
names. The flow of potential new drugs seemed fairly constant
over the years. (For the benefit of the statistician, assign-
ment of new USAN names is about 60 per year.) It was thus

obvious that the subject matter first addressed in Volume 1 was increasing at a fairly constant and impressive rate.

Once we had provided the background data up to 1976, it seemed logical to keep the series current by adding discussion of newer agents. Reports of drugs for new indications as well as the occurrence of brand-new structural types as drugs made it particularly important to update the existing volumes. The five-year cycle for preparation of new volumes represents a compromise between timeliness and comprehensiveness. A shorter period would date earlier entries. This volume thus covers compounds reported up to 1982.

As has been the practice in the earlier volumes, the only criterion for including a new therapeutic agent is its having been assigned a United States nonproprietary name (USAN), a so-called generic name. Since the focus of this text is chemistry, we have avoided in the main critical comments on pharmacology. The pharmacological activity or therapeutic utility described for the agents covered is that which was claimed when the USAN name was assigned.

The changes in chapter titles as well as changes in their relative sizes in going from volume to volume constitute an interesting guide to directions of research in medicinal chemistry. The first two volumes, for example, contained extensive details on steroid drugs. This section has shrunk to about a third of its former size in this book. The section on β-lactam antibiotics, on the other hand, has undergone steady growth from volume to volume: not only have the number of entries multiplied but the syntheses have become more complex.

This book, like its predecessors, is addressed to students
at the graduate level in organic and medicinal chemistry as
well as to practitioners in the field. It is again assumed
that the reader has a comfortable grasp of organic synthesis as
well as a basic grounding in biology.

We are pleased to acknowledge the helpful assistance of
several individuals in preparing this volume. Particularly, we
are grateful to Mrs. Janet Gill for preparing all of the
illustrations and to Mrs. Violet Huseby for long hours and
careful attention to detail in preparing the final copy and
several drafts.

Daniel Lednicer   Dublin, Ohio
Lester A. Mitscher   Lawrence, Kansas
                                                      January, 1984

Contents

THE ORGANIC CHEMISTRY
OF DRUG SYNTHESIS

VOLUME 3

1 Alicyclic and Cyclic Compounds

1. CYCLOPENTANES

a. Prostaglandins.

Few areas of organic medicinal chemistry in recent memory have had so many closely spaced pulses of intense research activity as the prostaglandins. Following closely on the heels of the discovery of the classical monocyclic prostaglandins (prostaglandin E_1, F_2, A_2, etc.), with their powerful associated activities, for example, oxytocic, blood pressure regulating, and inflammatory, was the discovery of the bicyclic analogues (the thromboxanes, prostacyclin) with their profound effects on hemodynamics and platelet function. More recently, the noncyclic leucotrienes, including the slow releasing substance of anaphylaxis, have been discovered. The activity these substances show in shock and asthma, for example, has excited considerable additional interest. Each of these discoveries has opened new physiological and therapeutic possibilites for exploitation. The newer compounds in particular are chemically and biologically short lived and are present in vanishingly small quantities so that much chemical effort has been expended on finding more efficient means of preparing them, on enhancing their stability, and on finding means of achieving greater tissue specificity.

In addition to its other properties, interest in the potential use of the vasodilative properties of prostaglandin E_1, alprostadil (4), has led to several conceptually different syntheses.[1-5] For this purpose, the classic Corey process[1] has to be modified by reversing the order of addition of the side chains to allow for convenient removal of the unwanted double bond in the upper side chain. For example, Corey lactone 1 is protected with dihydropyran (acid catalysis), reduced to the lactol with diisobutyaluminum hydride, and then subjected to the usual Wittig reaction to give intermediate 2. This is esterified with diazomethane, acetylated, and then catalytically hydrogenated to give intermediate 3 in which all of the oxygen atoms are differentiated. Further transformation to alprostadil (4) follows the well-trodden path of sequential Collins oxidation, Horner-Emmons olefination, zinc borohydride reduction, deetherification with aqueous acetic acid, separ-

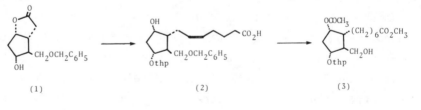

(1)   (2)   (3)

(4)

ation of the resulting C-15 epimers, dihydropyranylation, saponification of the ester groups, Jones oxidation (to introduce the C-9 keto group), and finally, deetherification.

The classic method for controlling stereochemistry is to perform reactions on cyclic substrates. A rather lengthy but nonetheless efficient example in the prostaglandin field uses bicyclic structures for this purpose.[2] Bisacetic acid derivative 5 is available in five steps from Diels-Alder reaction of trans-piperylene and maleic anhydride followed by side-chain homologation. Bromolactonization locks the molecule as bicyclic intermediate 6. Esterification, reductive dehalogenation (H_2/Raney Ni; $Cr(OAc)_2$), base opening of the lactone, careful esterification (CH_2N_2), and dehydration with methanesulfonyl chloride gives 7. The net result is movement of the double bond of 5. Treatment of 7 with NaH gives a fortunately unidirectional Dieckmann ring closure; alkylation with methyl ω-iodoheptanoate introduces the requisite saturated sidechain; lithium iodide-collidine treatment saponifies the ester during the course of which the extra carboxy group is lost; the sidechain methyl ester linkage is restored with diazomethane and the future keto group is protected by reaction with ethylene glycol and acid to give intermediate 8. Next, periodate-permanganate oxidation cleaves the double bond and leads to a methyl ketone whereupon the requisite trans-stereochemistry is established. Diazomethane esterification followed by Bayer-Villiger oxidation introduces the future C-11α hydroxyl group protected as the acetate. The dioxolane moiety at the future C-9 prevents β-elimination of the acetoxyl group of 9. In order to shorten the three-carbon sidechain, methoxide removes the acetyl group so that t-BuOK can close the lactone ring. NaH catalyzed condensation with methyl formate produces inter-

mediate <u>10</u>. Ozonization removes one carbon atom and acetic
anhydride is used to form enolacetate <u>11</u>, which intermediate is
now ready for excision of another carbon. Periodate-permanganate oxidation followed by ethylenediamine hydrolysis pro-
produces the needed aldehyde linkage, and the remainder of the
synthesis is rather straightforward. Horner-Emmons condensation produces ketone <u>12</u> which is sequentially protected with
trimethylsilyl chloride, and reduced with sodium borohydride,
the isomers separated, and then the blocking groups are removed
by base and then acid treatment to give <u>alprostadil(4)</u>.

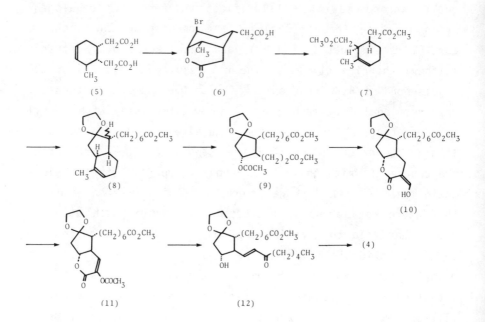

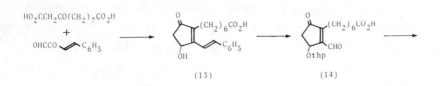

(13)   (14)

A conveniently short synthesis of __alprostadil__ begins with a mixed aldol assembly of the requisite cyclopentenone __13__.[3] This product is then oxidatively cleaved with periodate-permanganate and the alcohol moiety is protected as the tetrahydropyranyl ether (__14__). Aqueous chromous sulfate satisfactorily reduces the olefinic linkage and the __trans__ stereoisomer __15__ predominates after work-up. The remainder of the synthesis of __4__ involves the usual steps, through __16__ to __4__, with the exception that thexyl tetrahydrolimonyllithium borohydride is used to reduce the C-15 keto moiety so as to produce preferentially the desired C-15__S__ stereochemistry.

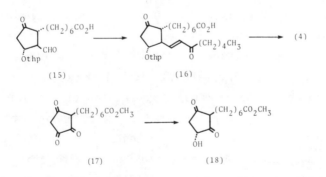

(15)   (16)

(17)   (18)

Consonant with the present interest in chiral synthesis, two additional contributions can be cited. Sih et al.[4] utilized a combined microbiological and organic chemical sequence in which key chirality establishing steps include the conversion of 17 to chiral, but unstable, 18 by enzymic reduction using the fungus Diplodascus uninucleatus. Lower side-chain synthon 20 was prepared by reduction of achiral 19 with Pencillium decumbens.

(19)   (20)

Stork and Takahashi[5] took D-glyceraldehyde synthon 21 from the chiral pool and condensed it with methyl oleate, using lithium diisopropylamide as catalyst for the mixed aldol reaction, leading to 22. The olefinic linkage is a latent form of the future carboxyl group. Protection of the diastereoisomeric mixture's hydroxyl by a methoxymethyleneoxo ether (MEMO) group and sequential acid treatments lead to β-lactone 23. This is tosylated, reduced to the lactol with dibal, and converted to the cyanohydrin (24). Ethylvinyl ether is used to cover the hydroxyl groups and then sodium hexamethyldisilazane treatment is used to express the nucleophilicity of the cyanohydrin ether, an umpohlung reagent for aldehydes that Stork has introduced. This internal displacement gives cyclopentane derivative 25. Periodate-permanganate oxidation cleaves the olefinic linkage, the ether groups are removed by dilute acid,

(21)   (22)

(23)   (24)

(25)   (26)

and diazomethane leads to the ester. The other protecting groups are removed to give chiral 26, which was already well known in its racemic form as a prostaglandin synthon.

A significant deactivating metabolic transformation of natural prostaglandins is enzymic oxidation of the C-15 hydroxyl to the corresponding ketone. This is prevented, with retention of activity, by methylation to give the C-15 tertiary carbinol series. This molecular feature is readily introduced at the stage of the Corey lactone (27) by reaction with methyl Grignard reagent or trimethylaluminum. The resulting mixture of tertiary carbinols (28) is transformed to oxytocic carba-prost (29) by standard transformations, including separation of diastereoisomers, so that the final product is the C-15 (R) analogue. This diastereoisomer is reputedly freer of typical prostaglandin side effects than the C-15 (S) isomer.[6]

Carbaprost can be converted to the metabolically stable

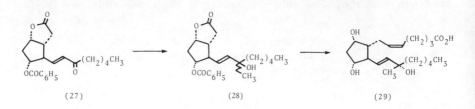

(27)   (28)   (29)

prostaglandin E analogue, arbaprostil (<u>31</u>), which exerts anti-secretory and cytoprotective activity in the stomach following oral administration and so promotes ulcer healing. At -45°C, selective silanization of the methyl ester of <u>carbaprost</u> gives <u>30</u>, which undergoes Collins oxidation and acid catalyzed de-blocking to produce <u>arbaprostil</u> (<u>31</u>).[6] The stereochemical configuration of the drug was confirmed by x-ray analysis. The branched alcoholic moiety can also be introduced by suitable modifications in the Horner-Emmons reaction.[7]

(29) ⟶

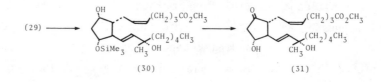

(30)   (31)

Another device for inhibiting transformation by lung pro-staglandin-15-dehydrogenase is introduction of <u>gem</u>-dimethyl branching at C-16. This stratagem was not sufficient, however, to provide simultaneously the necessary chemical stability to allow intravaginal administration in medicated devices for the purpose of inducing labor or abortion. It was found that this could be accomplished by replacement of the C-9 carbonyl group by a methylene (a carbon bioisostere) and that the resulting

agent, meteneprost (33), gave a lower incidence of undesirable
gastrointestinal side effects as compared with intramuscular
injection of carbaprost (29) methyl ester. The synthesis[8]
utilizes the sulfur ylide prepared from N,S-dimethyl-S-phenyl-
sulfoxime and methyl Grignard (32a). This reacts with 16,16-
dimethylprostaglandin E_2 methyl ester bis-(trimethylsilyl)
ester (32). The resulting β-hydroxysulfoximine undergoes olefination on reduction with aluminum amalgam[9] and deblocking
produces the uterine stimulant meteneprost (33).

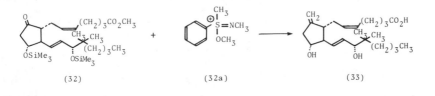

(32)   (32a)   (33)

Among the other metabolic transformations that result in
loss of prostaglandin activity is ω-chain oxidative degrada-
tion. A commonly employed device for countering this is to use
an aromatic ring to terminate the chain in place of the usual
aliphatic tail. Further, it is known in medicinal chemistry
that a methanesulfonimide moiety has nearly the same pK_a as a
carboxylic acid and occasionally is biologically acceptable as
well as a bioisostere. These features are combined in the
uterine stimulant, sulprostone (39). Gratifyingly these changes also result in both enhanced tissue selectivity toward the
uterus and lack of dehydration by the prostaglandin-15-dehydro-
genase.

The synthesis follows closely along normal prostaglandin

lines with the variations being highlighted here. Processed Corey lactone 34 undergoes Horner-Emmons trans olefination with ylide 35 to introduce the necessary features of the desired ω-side chain (36). After several standard steps, intermediate37 undergoes Wittig cis-olefination with reagent 38 and further standard prostaglandin transformations produce sulprostone (39).[10]

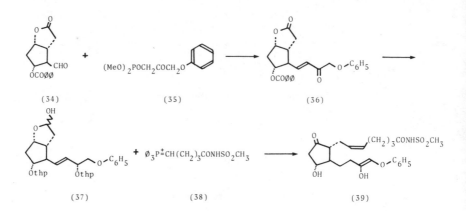

Thromboxane A_2, formed in blood platelets, is a vasocon-strictor with platelet aggregating action wheras prostacyclin, epoprostenol (43), formed in the lining cells of the blood vessels, is a vasodilator that inhibits platelet aggregation. Their biosynthesis from arachadonic acid via the prostaglandin cascade is normally in balance so that they together exert a sort of yin-yang balancing relationship fine tuning vascular homeostasis. The importance of this can hardly be overestimat-ed. Thrombosis causes considerable morbidity and mortality in advanced nations through heart attacks, stroke, pulmonary

embolism, thrombophlebitis, undesirable clotting associated
with implanted medical devices, and the like. Impairment of
vascular prostacyclin synthesis can well result in pathological
hypertension and excess tendency toward forming blood clots.
Administering exogenous prostacyclin, epoprostenol (43), shows
promise in combating these problems even though the drug is not
active if given orally and is both chemically and metabolically
unstable so that continuous infusion would seem to be needed
for normal maintenence therapy.

The drug is conveniently synthesized from prostaglandin
$F_{2\alpha}$ methyl ester (40), which undergoes oxybromination in the
presence of potassium triiodide to give 41. Treatment with DBN

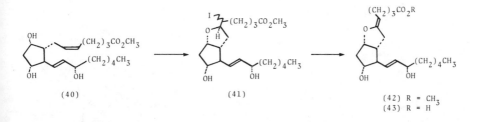

(40)   (41)   (42) R = CH_3
                                                            (43) R = H

(diazabicyclo[4.3.0]non-5-ene) gives dehydrohalogenation to
enol ether 42. Careful alkaline hydrolysis gives the sodium
salt of epoprostenol (43).[11] The free acid is extremely
unstable, presumably due to the expected acid lability of enol
ethers.

Much chemical attention is currently devoted to finding
chemically stable analogues of 43; Volume 4 will surely have
much to say about this.

   b. Retenoids
The discovery that some retinoids posess prophylactic activity
against carcinogenesis in epithelial tissues[12] has reawakened

interest in these terpene derivatives, particularly in 13-cis-retinoic acid (isotretinoin, 48) which is relatively potent and nontoxic. Isotretinoin also has keratolytic activity of value in the treatment of severe acne. The synthesis[13,14] is complicated by ready isomerization, and some early confusion existed in the literature regarding the identity of some intermediates. The natural terpene β-ionone (44) is subjected to a Reformatsky reaction with zinc and ethyl bromoacetate and the resulting product is reduced to the allylic alcohol with lithium aluminum hydride and then oxidized to trans-(β-ionylidene)acetaldehyde (45). This is condensed in pyridine with β-methylglutaconic anhydride to give 46. Careful saponification gives mainly diacid 47 which, on heating with copper and quinoline, decarboxylates to isotretinoin (48).[13,14]

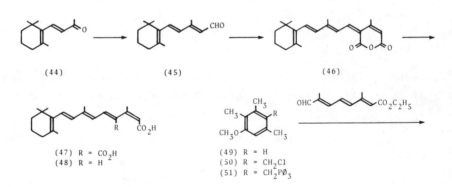

(44)   (45)   (46)

(47) R = CO$_2$H   (49) R = H
(48) R = H   (50) R = CH$_2$Cl
                           (51) R = CH$_2$PØ$_3$

The keratolytic analogue motretinide (53) is effective in treating acne and the excess epithelial growth characteristic of psoriasis, demonstrating that an aromatic terminal ring is compatible with activity. The synthesis[15] passes through the related orally active antipsoriatic/antitumor agent, etrinitate (52). These synthetic compounds have a wider safety margin than the natural materials. Etrinitate is synthesized[16] from 2,3,5-trimethylanisole by sequential chloromethylation (HCl and formaldehyde) to 50 followed by conversion to the ylid (51) with triphenylphosphine. Wittig olefination then leads to etrinitate (52). Etrinitate may then be saponified, activated by PCl_3 to the acid chloride, and then reacted with ethylamine to give motretinide (53).

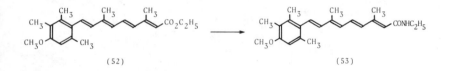

(52)   (53)

The retinoids share with certain steroid hormones the distinction of belonging to the few classes of substances capable of powerful positive influence on cell growth and differentiation.

    c. Miscellaneous

In building their characteristic cell walls, bacteria utilize D-alanine which they must manufacture enzymatically by epimerization of the common protein constituent, L-alanine, taken up in their diet. Because mammals have neither a cell wall nor an apparent need for D-alanine, this process is an attractive target for chemotherapists. Thus there has been developed a

group of mechanism-based inhibitors of alanine racemase. The
principle utilized in their design is that the enzyme would
convert an unnatural substrate of high affinity into a reactive
Michael acceptor which would then react with the enzyme to form
a covalent bond and inactivate the enzyme. Being unable to
biosynthesize an essential element of the cell wall, the organism so affected would not be able to grow or repair damage. It
was hypothesized that a strategically positioned halo atom
would eliminate readily in the intermediate pyridoxal complex
(54) to provide the necessary reactive species. A deuterium
atom at the α-carbon is used to adjust the rate of the process

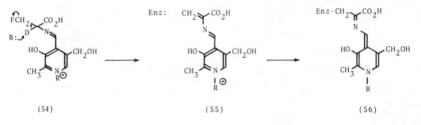

            (54)   (55)   (56)

so that the necessary reactions occur at what is judged to be
the best possible pace. The process is shown schematically
above (54 to 56) for the drug fludalanine (56). In practice,
the drug is combined with the 2,4-pentanedione enamine of cycloserine. The combination is synergistic as cycloserine inhibits the same enzyme, but by a different mechanism.

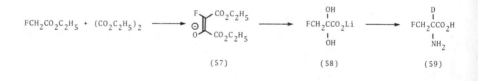

            (57)   (58)   (59)

One of the syntheses of fludalanine begins with base pro-
moted condensation of ethyl fluoroacetate and ethyl oxalate to
give 57. This is then converted by hydrolytic processes to the
insoluble hydrated lithium salt of fluoropyruvate (58). This
last is reductively aminated by reduction with sodium boro-
deuteride and the resulting racemate is resolved to give D-flu-
dalanine (59).[17]

There is a putative relationship between the pattern of
certain lipids in the bloodstream and pending cardiovascular
accidents. As a consequence, it has become a therapeutic ob-
jective to reduce the deposition of cholesterol esters in the
inner layers of the arterial wall. One attempts through diet
or the use of prophylactic drug treatments to reduce the amount
of very low density lipoproteins without interfering with high
density lipoproteins in the blood. The latter are believed to
be beneficial for they transport otherwise rather water insoluble cholesterol. Clofibrate, one of the main hypocholesterol-
emic drugs, has been shown to have unfortunate side effects in
some patients so alternatives have been sought. Gemcadiol
(62) is one of the possible replacements. This compound may be
synthesized by alkylating two molar equivalents of the cyclo-
hexylamine imine of isopropanal (60) with 1,6-dibromohexane
under the influence of lithium diisopropylamide. The resulting
dialdehyde (61) is reduced to gemcadiol (62) with sodium boro-

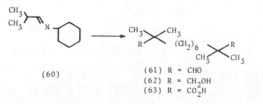

(60)

(61) R = CHO
(62) R = CH_2OH
(63) R = CO_2H

hydride.[18] There is evidence that gemcadiol is metabolically converted to diacid (63) which is believed to be the active agent at the cellular level.

REFERENCES

1. T. J. Schaff and E. J. Corey, J. Org. Chem., 37, 2921 (1972).

2. H. L. Slates, Z. S. Zelawski, D. Taub and N. L. Wendler, Tetrahedron, 30, 819 (1974).

3. M. Miyano and M. A. Stealey, J. Org. Chem., 40, 1748 (1975).

4. C. J. Sih, R. G. Salomon, P. Price, R. Sood and G. Peruzzotti, J. Am. Chem. Soc., 97, 857 (1975); C. J. Sih, J. B. Heather, R. Sood, P. Price, G. Peruzzotti, L. F. Hsu Lee, and S. S. Lee, ibid., 865.

5. G. Stork and T. Takahashi, J. Am. Chem. Soc., 99, 1275 (1977).

6. E. W. Yankee, U. Axen, and G. L. Bundy, J. Am. Chem. Soc., 96, 5865 (1974).

7. E. W. Yankee and G. L. Bundy, J. Am. Chem. Soc., 94, 3651 (1972); G. Bundy, F. Lincoln, N. Nelson, J. Pike, and W. Schneider, Ann. N.Y. Acad. Sci., 76, 180 (1971).

8. F. A. Kimball, G. L. Bundy, A. Robert, and J. R. Weeks, Prostaglandins, 17, 657 (1979).

9. C. R. Johnson, J. R. Shanklin, and R. A. Kirchoff, J. Am. Chem. Soc., 95, 6462 (1973).

10. T. K. Schaff, J. S. Bindra, J. F. Eggler, J. J. Plattner, J. A. Nelson, M. R. Johnson, J. W. Constantine, H.-J. Hess, and W. Elger, J. Med. Chem., 24, 1353 (1981).

11. R. A. Johnson, F. H. Lincoln, E. G. Nidy, W. P. Schneider, J. L. Thompson, and U. Axen, J. Am. Chem. Soc., 100, 7690 (1978).

12. D. L. Newton, W. R. Henderson, and M. B. Sporn, Cancer Res., 40, 3413 (1980).

13. C. D. Robeson, J. D. Cawley, L. Weister, M. H. Stern, C. C. Eddinger, and A. J. Chechak, J. Am. Chem. Soc., 77, 41111 (1955).

14. A. H. Lewin, M. G. Whaley, S. R. Parker, F. I. Carroll, and C. G. Moreland, J. Org. Chem., 47, 1799 (1982).

15. W. Bollag, R. Rueegg, and G. Ryser, Swiss Patent 616,134 (1980); Chem. Abstr., 93, 71312j (1980).

16. W. Bollag, R. Rueegg, and G. Ryser, Swiss Patent 616,135 (1980); Chem. Abstr., 93, 71314m (1980).

17. U.-H. Dolling, A. W. Douglas, E. J. J. Grabowski, E. F. Schoenewaldt, P. Sohar, and M. Sletzinger, J. Org. Chem., 43, 1634 (1978).

18. G. Moersch and P. L. Creger, U.S. Patent 3,929,897 (1975); Chem. Abstr., 85, 32426q (1976).

2 Phenethyl and Phenoxypropanolamines

The phenylethanolamine derivatives epinephrine (<u>1</u>) and nor-epinephrine (<u>2</u>) are intimately associated with the sympathetic nervous system. These two neurotransmitter hor-

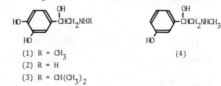

(1) R = CH₃
(2) R = H
(3) R = CH(CH₃)₂

(4)

mones control many of the responses of this branch of the involuntary, autonomic nervous system. Many of the familiar responses of the "fight or flight" syndrome such as vasoconstriction, increase in heart rate, and the like are mediated by these molecules. The profound biological effects elicited by these molecules have spurred an enormous amount of synthetic medicinal chemistry a better understanding of the action of the compounds at the molecular level and aimed also at producing new drugs. The availability of analogues of the natural substances interestingly led to the elucidation of many new pharmacological concepts. In spite of the fact that they differ only by an N-methyl group, the actions of epinephrine and norepinephrine are not quite the same. The former tends to elicit a largely inhibiting effect on most responses whereas the latter in general has a permissive action. These trends were accentuated in the close analogues isoproterenol (3) and phenylephrine (4). The pharmacology that lead to the division of the sympathetic nervous system into the α- and β-adrenergic branches was put on firmer footing by the availability of these two agents. It may be mentioned in passing that isoproterenol is an essentially pure β-adrenergic agonist whereas phenylephrine acts largely on the α-adrenergic system.

The search for new drugs in this series has concentrated quite closely on their action on the lungs, the heart and the vasculature. Medicinal chemists have thus sought sympathomimetic agents that would act exclusively as bronchodilating agents or as pure cardiostimulant drugs. The adventitious discovery that molecules which antagonize the action of β-sympathomimetic agents - the β-blockers - lower blood pressure has led to a corresponding effort in this field.

1. PHENYLETHANOLAMINES

As noted above, β-adrenergic agonists such as epinephrine typically cause relaxation of smooth muscle. This agent would thus in theory be useful as a bronchodilator for treatment of asthma; epinephrine itself, however, is too poorly absorbed orally and too rapidly metabolized to be used in therapy. A large number of analogues have been prepared over the years in attempts to overcome these short-comings. The initial strategy consisted in replacing the methyl group on nitrogen with an alkyl group more resistant to metabolic N-dealkylation. Isoproterenol (3) is thus one of the standbys as a drug for treatment of asthma.

The tertiary butyl analogue, colterol (9) is similarly resistant to metabolic inactivation. (It might be noted that there is some evidence that these more lipophilic alkyl groups, besides providing resistance to inactivation, also result in higher intrinsic activity by providing a better drug receptor interaction.) This drug can in principle be prepared by the scheme typical for phenylethanolamines. Thus acylation of catechol by means of Friedel-Crafts re-action with acetyl chloride affords the ketone 6; this is then halogenated to give intermediate 7. Displacement of bromine by means of tertiary butylamine gives the amino-ketone 8. Reduction of the carbonyl group by catalytic hydrogenation affords colterol (9).

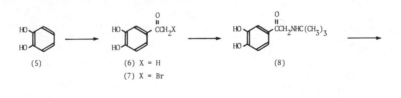

(5)   (6) X = H   (8)
             (7) X = Br

(9)

Absorption of organic compounds from the gastrointestinal tract is a highly complex process which involves at one one stage passage through a lipid membrane. Drugs that are highly hydrophilic thus tend to be absorbed very inefficiently by reason of their preferential partition into aqueous media. One strategy to overcome this unfavorable distribution consists in preparing a derivative that is more hydrophobic and which will revert to the parent drug on exposure to metabolizing enzymes after absorption. Such derivatives, often called prodrugs, have been investigated at some length in order to improve the absorption characteristics of the very hydrophilic catecholamines.

Acylation of aminoketone <u>8</u> with the acid chloride from p-toluic acid affords the corresponding ester (<u>10</u>); catalytic hydrogenation leads to the bronchodilator <u>bitolerol</u> (<u>11</u>)[1]. An analogous scheme starting from the N-methyl ketone (<u>12</u>) and pivaloyl chloride gives aminoalcohol (<u>14</u>). This compound is then resolved to isolate the levorotatory isomer[2]. There is thus obtained the drug <u>dipivefrin</u>.

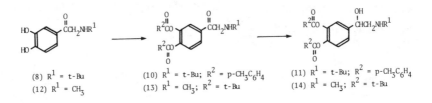

(8) R^1 = t-Bu
(12) R^1 = CH_3

(10) R^1 = t-Bu; R^2 = p-$CH_3C_6H_4$
(13) R^1 = CH_3; R^2 = t-Bu

(11) R^1 = t-Bu; R^2 = p-$CH_3C_6H_4$
(14) R^1 = CH_3; R^2 = t-Bu

A variant on this theme contains mixed acyl groups. In the absence of a specific reference it may be speculated that the synthesis starts with the diacetyl derivative (15). Controlled hydrolysis would probably give the monoacetate (16) since the ester para to the ketone should be activated by that carbonyl function. Acylation with anisoyl chloride followed by reduction would then afford nisobuterol (18).

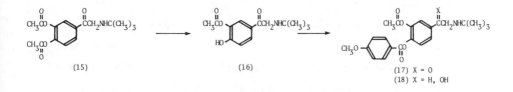

Catecholamines are also intimately involved in cardiac function, with β-sympathetic agonists having a generally stimulant action on the heart. Some effort has thus been devoted to the synthesis of agents that would act selectively on the heart. (Very roughly speaking, β^1-adrenergic receptor agonists tend to act on the heart while β^2-adrenergic receptor agonists act on the lungs; much the same holds true for antagonists; see below.)

Preparation of the cardiotonic agent butopamine (23) starts with reductive amination of ketone 19. Acylation of the resulting amide (20) with hydroxyacid 21 affords the corresponding amine (22). Treatment with lithium aluminum hydride serves both to reduce the amide and remove the acetyl protecting groups. There is thus obtained butopamine [3].

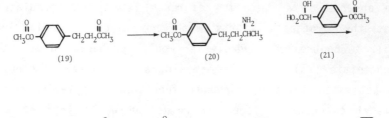

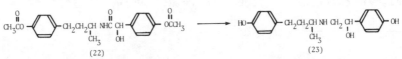

Drugs that block the action of α-adrenergic activation effectively lower blood pressure by opposing the vasoconstricting effects of norepinephrine. Drawbacks of these agents, which include acceleration of heart rate, orthostatic hypotension and fluid retention, were at one time considered to be due to the extension of the pharmacology of α-blockers. Incorporation of β-blocking activity into the molecule should oppose these effects. This strategy seemed particularly promising in view of the fact that β-adrenergic blockers were adventitiously found lower blood pressure in their own right. The first such combined α- and β-blocker, labetolol has confirmed this strategy and proved to be a clinically useful antihypertensive agent.

The drugs in this class share the phenylethanolamine moiety and a catechol surrogate in which the 3-hydroxyl is replaced by some other function that contains relatively acidic protons.

Synthesis of the prototype[4] begins with Friedel Crafts acetylation of salicylamide (24). Bromination of the ketone (25) followed by displacement with amine 27 gives the corresponding aminoketone (28). Catalytic hydrogenation to the aminoalcohol completes the synthesis of labetolol (24). The presence of two chiral centers at remote positions leads to the two diastereomers being obtained in essentially equal amounts.

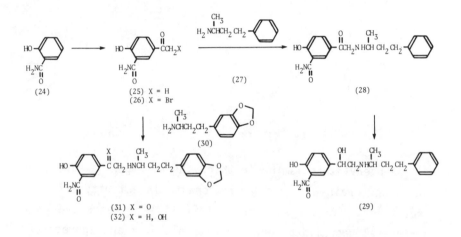

In much the same vein, alkylation of bromoketone (26) with amine (30) (obtained by reductive amination of the corresponding ketone) affords aminoketone (31). Catalytic reduction leads to medroxalol (32)[5].

The methyl group on a sulfoxide interestingly proves sufficiently acidic to substitute for phenolic hydroxyl. The preparation of this combined α- and β-blocker, sulfinalol[6], begins by protection of the phenolic hydroxyl as its benzoate ester (34). Bromination (35) followed by

condensation with amine <u>36</u> gives the aminoketone (<u>37</u>).
Successive catalytic reduction and saponification affords
the aminoalcohol (<u>38</u>).   Oxidation of the sulfide to he
sulfoxide with a reagent such as metaperiodate gives
<u>sulfinalol</u> (<u>39</u>).   This last step introduces a third chiral
center because trigonal sulfur exists in antipodal forms.
The number of diastereomers is thus increased to eight.

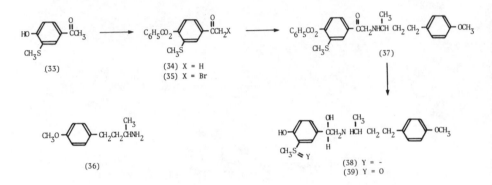

(33)

(34) X = H
(35) X = Br

(37)

(36)

(38) Y = -
(39) Y = O

A phenylethanolamine in which the nitrogen is alkylated
by a long chain alphatic group departs in activity from the
prototypes.   This agent, <u>suloctidil</u> (<u>43</u>) is described as a
peripheral vasodilator endowed with platelet antiaggregatory
activity. As with the more classical compunds, preparation
proceeds through bromination of the substituted propiophen-
one (<u>40</u>) and displacement of halogen with octylamine.   Reduction, in this case by means of sodium borohydride affords
<u>suloctidil</u> (<u>43</u>)[7].

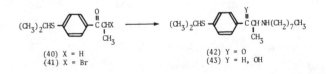

(40) X = H
(41) X = Br

(42) Y = O
(43) Y = H, OH

Pharmacological theory would predict that β-adrenergic blockers should oppose the vasodilating action of epinephrine and, in consequence, increase blood pressure. It was found, however, that these drugs in fact actually decrease blood pressure in hypertensive individuals, by some as yet undefined mechanism. The fact that this class of drugs tends to be very well tolerated has led to enormous emphasis on the synthesis of novel β-blockers. The observation that early analogues tended to exacerbate asthma by their blockade of endogenous β-agonists has led to the search for compounds that show a preference for β^1-adrenergic sites.

With some important exceptions, drugs in this class are conceptually related to the phenylethanolamines by the interposition of an oxymethylene group between the aromatic ring and the benzyl alcohol.

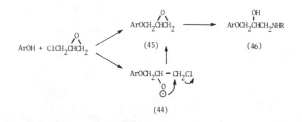

Compounds are prepared by a fairly standard sequence which consists of condensation of an appropriate phenol with epichlorohydrin in the presence of base. Attack of phenoxide can proceed by means of displacement of chlorine to give epoxide (45) directly. Alternatively, opening of the epoxide leads to anion 44; this last, then, displaces halogen on the adjacent carbon to lead to the same epoxide. Reaction of the epoxide with the appropriate amine then completes the synthesis.

Application of this scheme to o-cyclopentyl phenol, o-cyclohexylphenol and m-cresol thus leads to respectively, penbutolol (47)[8], exapralol (48)[9] and bevantolol (49) [10]. The phenoxypanolamine tipropidil (52) interestingly exhibits much the same biological activity as its phenylethanolamine parent suloctidil (53).

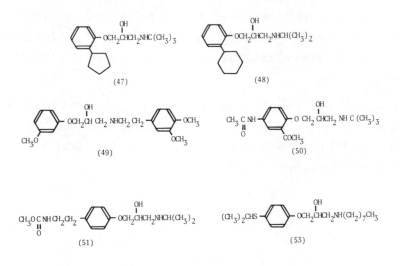

(47)

(48)

(49)

(50)

(51)

(53)

The phenol (55) required for preparation of diacetolol (50)[11] can be otained by Friedel-Crafts acetylation of p-acetamidophenol. The starting material (58) for pamatolol (51)[12] can be derived from p-hydroxyphenylacetonitrile (56) by reduction to the amine (57) followed by treatment with ethyl chloroformate. Bucindolol (52) is one of the newer β-blockers designed to incorporate non-adrenergically mediated vasodilating activity in the same molecule as the adrenergic blocker. Preparation of the amine (61) for this

agent starts by displacement of the dimethylamino group in
gramine (59) by the anion from 2-nitropropane. Reduction of
the nitro group leads to the requisite intermediate[13].

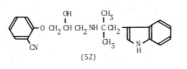

(52)

(54)   (55)

Synthesis of primidolol (65)[14] can be carried out by a
convergent scheme. One branch consists in application of
the usual scheme to o-cresol (62); ring opening of the
intermediate oxirane with ammonia leads to the primary amine
(63). The side chain fragment (64) can be prepared by
alkylation of pyrimidone (63) with ethylene dibromide to
afford 64. Alkylation of aminoalcohol 62 with halide 64
affords primidolol.

NC CH₂ ⬡ OH ⟶ H₂ NCH₂ CH₂ ⬡ OH ⟶ CH₃ O C NH CH₂CH₂ ⬡ OH

(56)   (57)   (58)

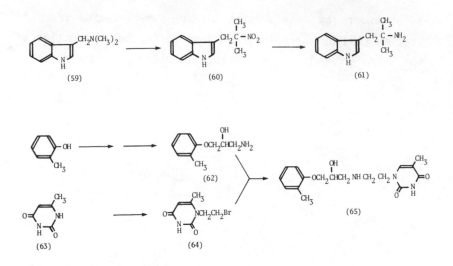

It is by now well accepted that most drugs, particularly those whose structures bear some relation to endogenous agonists owe their effects to interaction with bio-
polymer receptors. Since the latter are constructed from
chiral subunits (amino acids, sugars, etc.), it should not
be surprising to note that drugs too show stereoselectivity
in their activity. That is, one antipode is almost invariably more potent than the other. In the case of the
adrenergic agonists and antagonists, activity is generally
associated with the R isomer. Though the drugs are, as a
rule, used as racemates, occasional entities consist of
single enantiomers. Sereospecific synthesis is, of course,
preferred to resolution since it does not entail discarding
half the product at the end of the scheme.

Prenalterol (73) interestingly exhibits adrenergic
agonist activity in spite of an interposed oxymethylene
group. The stereospecific synthesis devised for this
molecule[15] relies on the fact that the side chain is very

similar in oxidation state to that of a sugar. Condensation of the monobenzyl ether of phenol 66 with the epoxide derived from D-glucofuranose (67) affords the glycosylated derivative (68). Hydrolytic removal of the protecting groups followed by cleavage of the sugar with periodate gives aldehyde 69. This is in turn reduced to the glycol by means of sodium borohydride and the terminal alcohol is converted to the mesylate (71). Displacement of that group with isopropylamine (72) followed by hydrogenolytic removal of the O-benzyl ether affords the β^2 - selective adrenergic agonist prenalterol (73).

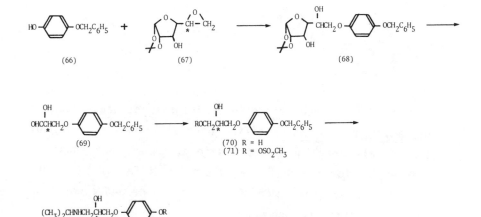

Formal cyclization of the hydroxyl and amine functions to form a morpholine interestingly changes biological activity markedly; the resulting compound shows CNS activity as an antidepressant rather than as an adrenegic agent. Reaction of epoxide (74) with the mesylate from ethanolamine leads to viloxazine (76) in a single step[16]. It is likely that reaction is initiated by opening of the oxirane by the amino group. Internal displacement of the leaving group by the resulting alkoxide forms the morpholine ring.

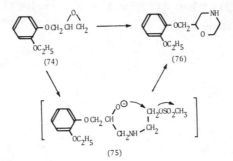

The widely used tricyclic antidepressant drugs such as imipramine and amitriptyltiline have in common a series of side effects that limit their safety. There has thus occasioned a wide search for agents that differ in structure and act by some other mechanism. Nisoxetine and fluoxetine are two nontricyclic compounds which have shown promising early results as antidepressants. Mannich reaction on acetophenone leads to the corresponding aminoketone (78). Reduction of the carbonyl group (79) followed by replacement of the hydroxyl by chlorine gives intermediate 80. Displacement of chlorine with the alkoxide from the monomethyl ether of catechol gives the corresponding aryl ether (81). The amine is then dealkylated to the monomethyl derivative by the von Braun sequence (cyanogen bromide followed by base) to give nisoxetine (82). Displacement on (80) with the monotrifluoromethyl ether from hydroquinone followed by demethylation leads to fluoxetine (84)[17].

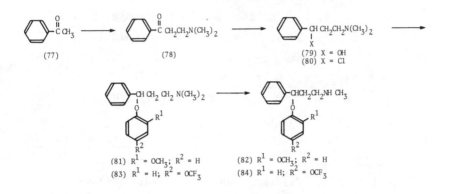

(77) (78)

(79) X = OH
(80) X = Cl

(81) R^1 = OCH_3; R^2 = H   (82) R^1 = OCH_3; R^2 = H
(83) R^1 = H; R^2 = OCF_3 (84) R^1 = H; R^2 = OCF_3

REFERENCES

1. M. Minatoya B. F. Tullar and W. D. Conway, U.S. Patent 3,904,671; Chem. Abstr. 84, 16943e (1976).
2. A. Hussain and J. E. Truelove, German Offen. 2,343,657; Chem. Abstr. 80, 145839s (1974).
3. J. Mills, K. K. Schmiegel and R. R. Tuttle, Eur. Patent Appl. 7,205 (1980); Chem. Abstr. 93, 94972 (1980).
4. L. H. C. Lunts and D. T. Collin, German Offen. 2,032,642; Chem. Abstr. 75, 5520c (1971).
5. J. T. Suh and T. M. Bare, U.S. Patent 3,883,560; Chem. Abstr. 83, 78914j (1975).
6. Anon.   British Patent 1,544,872; Chem. Abstr. 92, 163686s (1980).
7. G. Lambelin, J. Roba, and C. Gillet, German Offen. 2,344,404; Chem. Abstr. 83, 97820 (1975).
8. G. Haertfelder, H. Lessenich and K. Schmitt, Arzneim. Forsch. 22, 930 (1972).
9. M. Carissimi, P. Gentili, E. Grumelli, E. Milla, G. Picciola and F. Ravenna, Arzneim. Forsch. 26, 506 (1976).
10. M. Ikezaki, K, Irie, T. Nagao, and K. Yamashita, Japanese Patent 77, 00234; Chem. Abstr. 86, 1894767 (1977).
11. K. R. H. Wooldridge and B. Berkley, South African Patent 68 03,130; Chem. Abstr. 70, 114824 (1969).
12. A. E. Brandstrom, P. A. E. Carlsson, H. R. Corrodi, L. Ek and B. A. H. Ablad, U.S. Patent 3,928,601; Chem. Abstr. 85, 5355j (1976).

13. W. E. Kreighbaum, W. L. Matier, R. D. Dennis, J. L. Minielli, D. Deitchman, J. L. Perhach, Jr. and W. T. Comer, J. Med. Chem., 23, 285 (1980).

14. J. Augstein, D. A. Cox and A. L. Ham., German Offen. 2,238,504 (1973). Chem. Abstr. 78, 136325e (1973).

15. K. A. Jaeggi, H. Schroeter, and F. Ostermayer, German Offen. 2,503,968; Chem. Abstr. 84, 5322 (1976).

16. S. A. Lee, British Patent 1,260,886; Chem. Abstr. 76, 99684e (1972).

17. B. B. Molloy and K. K. Schmiegel, German Offen. 2,500,110; Chem. Abstr. 83, 192809d (1975).

3 Arylaliphatic Compounds

The aromatic portion of the molecules discussed in this chapter is frequently, if not always, an essential contributor to the intensity of their pharmacological action. It is, however, usually the aliphatic portion that determines the nature of that action. Thus it is a common observation in the practice of medicinal chemistry that optimization of potency in these drug classes requires careful attention to the correct spatial orientation of the functional groups, their overall electronic densities, and the contribution that they make to the molecule's solubility in biological fluids. These factors are most conveniently adjusted by altering the substituents on the aromatic ring.

1. ARYLACETIC ACID DERIVATIVES

The potent antiinflammatory action exerted by many arylacetic acid derivatives has led to the continued exploration of this class. It is apparent from a consideration of the structures of compounds that have become prominent that considerable structural latitude is possible without loss of activity.

The synthesis of fenclofenac (5), a nonsteroidal antiinflammatory agent (NSAI), starts with condensation of o-chloro-

37

acetophenone (<u>1</u>) and 2,4-dichlorophenol (<u>2</u>) under Ullmann conditions (Cu/NaOH). The unsymmetrical diarylether (<u>3</u>) is subjected to the Willgerodt-Kindler reaction to give thioamide <u>4</u>. This last is saponified to produce <u>fenclofenac</u> (<u>5</u>).[1]

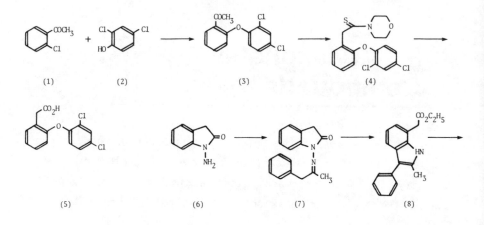

A structure more distantly related to these is <u>amfenac</u> (<u>10</u>). Like most of the others, <u>amfenac</u>, frequently used after tooth extraction, is an antiinflammatory agent by virtue

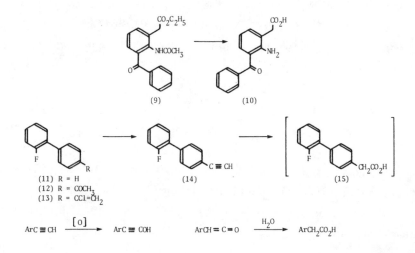

of inhibition of the cyclooxygenase enzyme essential for prostaglandin biosynthesis. The synthesis begins with hydrazone
(7) formation between phenylacetone and 1-aminoindolin-2-one
(6) by warming in acetic acid. Treatment with HCl/EtOH results
in a Fischer indole rearrangement to produce 8. Ozonolysis
produces unsymmetrical diarylketone 9 which is converted to an
intermediate indolin-2-one by cleavage of the ester and acetamide moieties with HCl and then lactam hydrolysis with NaOH to
give amfenac (10).[2]

A closet member of this little group, revelation of whose
real nature requires metabolic transformation of an acetylenic
linkage to an acetic acid moiety, is fluretofen (14). The
synthesis begins by Friedel-Crafts acylation of 2-fluorobi-
phenyl (11) with acetic acid to give ketone 12. Heating with
PCl_5/$POCl_3$ then produces the α-chlorostyrene analogue 13.
Dehydrohalogenation with strong base ($NaNH_2$) produces the
acetylene, fluretofen (14). Metabolic studies reveal facile
bioconversion to the arylacetic acid analogue 15. This is an
active antiinflammatory agent when administered directly and so
is believed to account for the activity of fluretofen. A likely pathway for this transformation involves terminal oxygen-

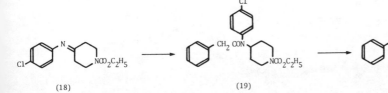

ation of the acetylenic moiety, which product would then tauto-
merize to the ketene. Spontaneous hydration of the latter
would complete the sequence.[3]

Phenylacetamides have a variety of pharmacological actions
depending upon the nature of the amine-derived component.

Guanfacine (17), an antihypertensive agent acting as a
central α-adrenergic receptor agonist, requires administration
only once daily and reportedly has fewer CNS sideeffects than
the somewhat related drug, clonidine. Guanfacine is prepared
readily by ester-amide exchange of methyl 2,6-dichlorophenyl
acetate (16) using guanidine.[4]

Use of a large, lipophilic nitrogenous component results
in a lidocaine like, local anesthetic type cardiac antiarrhythmic drug, lorcainide (20). Synthesis begins with the
Schiff's base (18) derived by reaction of p-chloro-aniline and
borohydride followed by acylation with phenylacetyl chloride
produces amide 19 .Selective hydrolysis with HBr followed by
alkylation with isopropyl bromide completes the synthesis
of lorcainide (20).[5]

The structural requirements for such activity are not very
confining, as can be seen in part by comparing the structure
of lorcainide with oxiramide (21). Antiarrhythmic oxiramide
(21) is made by a straightforward ester-amide exchange reaction

(21)

involving ethyl 2-phenoxyphenylacetate and 4-cis-(2,6-dimethyl-piperidino)butylamine.[6]

Introducing yet more structural complexity into the amine component leads to the antiarrhythmic agent disobutamide (underline{24}). Disobutamide is structurally related to disopyramide (underline{25})[7] but is faster and longer acting. The synthesis of underline{24} begins with sodamide induced alkylation of 2-chlorophenylacetonitrile with 2-diisopropylaminoethyl chloride to give underline{22}. A second sodamide mediated alkylation, this time with 2-(1-piperidino)ethyl chloride, gives nitrile underline{23}. Subsequent sulfuric acid hydration completes the synthesis of disobutamide (underline{24}).[8]

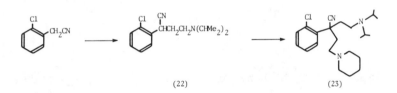

(22)   (23)

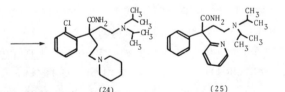

(24)   (25)

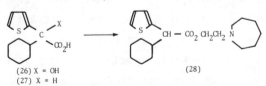

(26) X = OH   (28)
(27) X = H

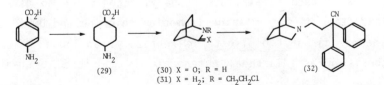

(29)   (30) X = O; R = H   (32)
            (31) X = H₂; R = CH₂CH₂Cl

The treacheries inherent in naive attempts at pattern recognition are illustrated by the finding that ester 28, known as cetiedil, is said to be a peripheral vasodilator. Clemmensen reduction of Grignard product 26 removes the superfluous benzylic hydroxyl group and esterification of the sodium salt of the resulting acid (27) with 2-(1-cycloheptylamino)ethyl chloride produces cetiedil (28).[9]

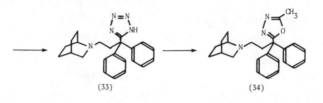

(33)   (34)

An intresting biphenyl derivative utilizing a bioisosteric replacement for a carboxyl group is the antidiarrheal agent, nufenoxole (34). To get around addictive and analgesic side effects associated with the classical morphine based antidiarrheal agents, a different class of drug was sought. Nufenoxole has few analgesic, anticholinergic, or central effects. Reduction with a ruthenium catalyst (to prevent hydrogenolysis) converts p-aminobenzoic acid to cyclohexane derivative 29. Internal N-acetylation of the cis isomer followed by heating gives bicyclic lactam 30. Hydride reduction to the isoquineuclidine and alkylation gives 2-azabicyclo[2.2.2]octane synthon 31. This is used to alkylate diphenylacetonitrile to give 32. Cycloaddition of sodium azide (ammonium chloride and DMF) gives the normal carboxyl bioisosteric tetrazolyl analogue 33. The synthesis of antidiarrheal nufenoxole is completed by heating

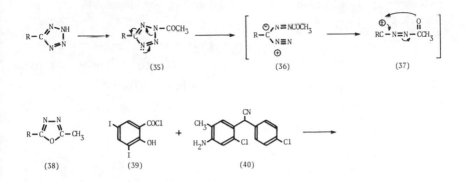

(35)   (36)   (37)

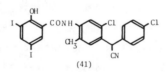

(38)   (39)   (40)

with acetic anhydride to give the 2-methyl-1,3,4-oxadiazol-5-yl analogue.[10,11] The mechanism of this rearrangement is believed to involve N-acetylation (35) with subsequent ring opening to the diazoalkane (36) which loses nitrogen to give carbene 37, which cyclizes to the oxadiazole (38).[12]

(41)

(42) X = OH
(43) X = Cl
(44) X = NHC$_2$H$_5$
(45) X = NH$_2$

(46) X = NH$_2$
(47) X = OH

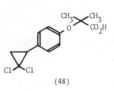

(48)

A phenylacetonitrile derivative, closantel (41), is an anthelmintic agent useful against sheep liver flukes. Its patented synthesis involves a Schotten-Baumann amidation between acid chloride 39 and complex aniline 40 to give closantel (41).[13]

A cinnamoylamide, cinromide (44), is a long-acting anti-convulsant similar in its clinical effects to phenacetamide but is less hepatotoxic. The synthesis involves the straight-forward amidation of acid 42 via the intermediate acid chloride (SOCl$_2$) 43. It appears that the drug is mainly deethylated in vivo to give active amide 45.[14]

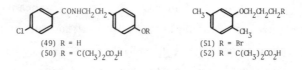

(49) R = H
(50) R = C(CH$_3$)$_2$CO$_2$H

(51) R = Br
(52) R = C(CH$_3$)$_2$CO$_2$H

2,2-Disubstituted aryloxyacetic acid derivatives related to clofibrate have been intensively studied in an attempt to get around the side effects of the latter drug.

Ciprofibrate (48), a more potent lipid-lowering agent than clofibrate, is prepared from Simmons-Smith product 46 by Sandmeyer replacement of the amino group by a hydroxyl via the diazonium salt. Phenol 47 undergoes the Reimer-Thiemann like process common to these agents upon alkaline treatment with acetone and chloroform to complete the synthesis of ciprofib-rate (48).[15]

Further indication that substantial bulk tolerance is available in the para position is given by the lipid lowering agent bezafibrate (50). The p-chlorobenzamide of tyramine (49) undergoes a Williamson ether synthesis with ethyl 2-bromo-

7-methylpropionate to complete the synthesis. The ester group
is hydrolyzed in the alkaline reaction medium.[16]

Apparently a substantial spacer is also allowable between
the aromatic ring and the carboxy group. Gemfibrozil (52), a
hypotriglyceridemic agent which decreases the influx of steroid
into the liver, is a clofibrate homologue. It is made readily
by lithium diisopropylamide-promoted alkylation of sodium iso-
propionate with alkyl bromide 51.[17]

A rather distantly related analogue incorporating a β-di-
carbonyl moiety as a bioisosteric replacement for a carboxyl,
arildone (55), blocks the uncoating of polio virus and herpes
simplex virus type I and thus inhibits infection of cells and
the early stages of virus replication. Thus effective therapy
would require careful timing as it does with amantidine.
Alkylation of phenol 53 with 1,6-dibromohexane gives haloether

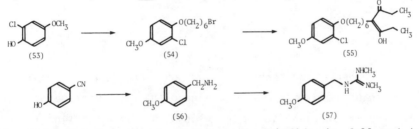

54. Finkelstein reaction with sodium iodide is followed by
acylation of heptane-3,5-dione to complete the synthesis of
arildone (55).[18]

2. ANILINES, BENZYL AMINES, AND ANALOGUES
An orally active local anesthetic agent that can be used as an
antiarrhythmic agent is meobentine (57). Its patented synthesis starts with p-hydroxyphenylnitrile and proceeds by dimethyl
sulfate etherification and Raney nickel reduction to 56.
Alkylation of S-methyl-N,N'-dimethylthiourea with 56 completes
the synthesis of meobentine (57).[19]

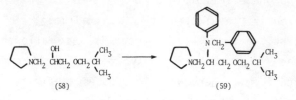

(58)   (59)

Bepridil (59) blocks the slow calcium channel and serves
as an antianginal agent and a vasodilator. In its synthesis,
alcohol 58 (derived from epichlorohydrin) is converted to the
corresponding chloride with thionyl chloride and displaced with
the sodium salt of N-benzylaniline to give bepridil (59)[20]

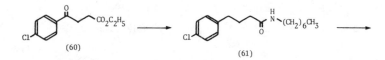

(60)   (61)

A number of quaternary amines are effective at modulating
nerve transmissions. They often have the disadvantage of being
relatively nonselective and so possess numerous sideeffects.
This contrasts with the advantage that they do not cross the
blood-brain barrier and so have no central sideeffects. Clofilium phosphate (63) is such an antiarrhythmic agent. It is
synthesized from ester 60 by saponification followed by Clemmensen reduction and amide formation (oxalyl chloride followed
by n-heptylamine) to give 61. Diborane reduction gives secondary amine 62. Reaction with acetyl chloride followed by another diborane reduction gives the tertiary amine. Finally, reaction with ethyl bromide and ion exchange with phosphate complete the synthesis of clofilium phosphate (63).[21]

(62)   (63)

Another quaternary antiarrhythmic agent is <u>emilium tosyl-</u><u>ate</u> (<u>65</u>). It is synthesized simply by quaternization of <u>m</u>-methoxybenzyl chloride (<u>64</u>) with dimethylethylamine followed by ion exchange.[22]

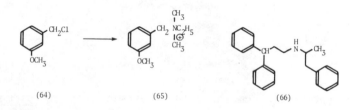

(64)   (65)   (66)

3. DIARYLMETHANE ANALOGUES

<u>Prenylamine</u> (<u>66</u>) was long used in the treatment of angina pect-oris, in which condition it was believed to act by inhibiting the uptake and storage of catecholamines in heart tissue. <u>Droprenilamine</u> (<u>69</u>), an analogue in which the phenyl ring is reduced, acts as a coronary vasodilator. One of several syn-theses involves simple reductive alkylation of 1,1-diphenyl-propylamine (<u>67</u>) with cyclohexylacetone (<u>68</u>).[23]

<u>Drobuline</u> (<u>71</u>) is a somewhat related cardiac-directed drug with antiarrhythmic action. Since both enantiomers have the

same activity, it is likely that its pharmacological action is due to a local anesthetic-like action. It is synthesized by sodium amide mediated alkylation of diphenylmethane with allyl bromide to give 70. Epoxidation with m-chloroperbenzoic acid followed by opening of the oxirane ring at the least hindered carbon by isopropylamine completes the synthesis.[24]

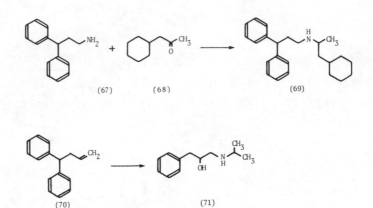

(67)   (68)   (69)

(70)   (71)

A slightly more complex antiarrhythmic agent is <u>pirmentol</u> (<u>74</u>). It is synthesized from 4-chloropropiophenone (<u>72</u>) by keto group protection as the dioxolane (with ethylene glycol and acid) followed by sodium iodide-mediated alkylation with <u>cis</u> 2,6-dimethylpiperidine to give 73. Deblocking with acid followed by addition of 2-lithiopyridine completes the synthesis of <u>pirmentol</u> (<u>74</u>).[25]

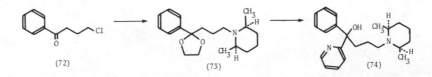

(72)   (73)   (74)

For many years after the discovery of the antidepressant activity of phenothiazine, almost all synthetic activity centered about rigid analogues. Recently attention has been paid to less rigid molecules in part because of the finding that zimelidine (77) is an antidepressant showing selective inhibition of the central uptake of 5-hydroxytryptamine and that it possesses less anticholinergic activity than amitriptylene. One of a number of syntheses starts with p-bromoacetophenone and a Mannich reaction (formaldehyde and dimethylamine) to give aminoketone 75. Reaction with 3-lithio-pyridine gives tertiary carbinol 76. Dehydration with sulfuric acid gives a mixture of Z and E forms of which the Z analogue is the more active.[26]

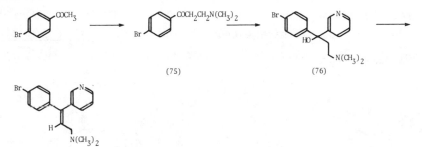

(75)   (76)

(77)

Pridefine (80) is a somewhat structurally related antidepressant. It is a centrally active neurotransmitter blocking agent. It blocks norepinephrine in the hypothalamus but does not affect dopamine or 5-hydroxytryptamine. Its synthesis begins by lithium amide-promoted condensation of diethyl succinate and benzophenone followed by saponification to 78. Heating in the presence of ethylamine gives N-ethylsuccinimide 79. Lithium aluminum hydride reduction completes the synthesis of pridefine (80).[27]

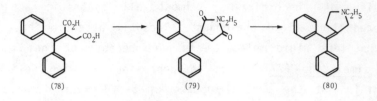

(78)   (79)   (80)

4. STILBENE ANALOGUES

Cells from tissues associated with primary and secondary sexual
characteristics are under particular endocrine control.   Sex
hormones determinethe growth, differentiation, and prolifer-
ation of such cells.   When a tumor develops in such tissues, it
is sometimes hormone dependent and the use of antihormones removes the impetus for the tumor's headlong growth.   Many nonsteroidal compounds have estrogenic activity; diethylstilbest-
irol (81) may be taken as an example.   Certain more bulky an-

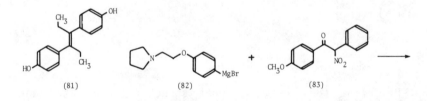

(81)   (82)   +   (83)

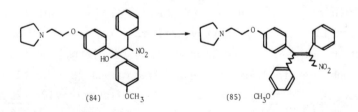

(84)   OCH₃   (85)   CH₃O

alogues are antagonists at the estrogenic receptor level and exert a second order anti-tumor response.

Nitromifene (85) is such an agent. A Grignard reaction of arylether 82 and ketone 83 leads to tertiary carbinol 84. Tosic acid dehydration leads to a mixture of Z and E stilbenes which constitute the antiestrogen, nitromifene (85).[28]

Another example is tamoxifen (89). Its synthesis begins with Grignard addition of reagent 86 to aryl ketone 87 giving carbinol 88. Dehydration leads to the readily separable Z and E analogues of 89. Interestingly, in rats the Z form is an antiestrogen whereas the E form is estrogenic. Metabolism involves p-hydroxylation and this metabolite (90) is more potent than tamoxifen itself. In fact, metabolite 90 may be the active form of tamoxifen (89).[29]

(86)   (87)   (88)

(89) X = H
(90) X = OH

REFERENCES

1. D. C. Atkinson, K. E. Godfrey, B. J. Jordan, E. C. Leach,
    B. Meek, J. D. Nichols, and J. F. Saville, J. Pharm.
    Pharmacol., 26, 357 (1974).

2. W. J. Welstead, Jr., H. W. Moran, H. F. Stauffer, L. B.
    Turnbull, and L. F. Sancillo, J. Med. Chem., 22, 1074
    (1979).

3. W. B. Lacefield and W. S. Marshall, U.S. Patent 3,928,604;
    H. R. Sullivan, P. Roffey, and R. E. McMahon, Drug Metab.
    Disposn., 7, 76 (1979).

4. J. B. Bream, H. Lauener, C. W. Picard, G. Scholtysik, and
    T. G. White, Arzneim. Forsch., 25, 1477 (1975).

5. H. K. F. Hermans and S. Sanczuk, U.S. Patent 4,197,303
    (1975); Chem. Abstr., 93, 132380d (1980).

6. Anon., Netherlands Patent, 6,605,452 (1962); Chem. Abstr.,
    66, 104914e (1967).

7. D. Lednicer and L. A. Mitscher, The Organic Chemistry of
    Drug Synthesis, Vol. 2, Wiley, New York, 1980, p. 81.

8. P. K. Youan, R. L. Novotney, C. M. Woo, K. A. Prodan, and
    F. M. Herschenson, J. Med. Chem., 23, 1102 (1980).

9. M. Robba and Y. LeGuen, Eur. J. Med. Chem., 2, 120 (1967).

10. G. W. Adelstein, C. H. Yen, E. Z. Dajani, and R. G.
    Biandi, J. Med. Chem., 19, 1221 (1976).

11. W. Schneider and R. Dillman, Chem. Ber., 96, 2377 (1963).

12. R. Huisgen, J. Sauer, H. J. Sturm, and J. H. Markgraf,
    Chem. Ber., 93, 2106 (1960).

13. M. A. C. Janssen and V. K. Sipido, German Offen.,
    2,610,837 (1976); Chem. Abstr., 86, 55186w (1977).

14. E. M. Grivsky, German Offen., 2,535,599 (1976); Chem.
    Abstr., 84, 164492x (1976).

15. D. K. Phillips, German Offen., 2,343,606 (1974); Chem.

Abstr, 80, 133048v (1974).

16. E. C. Witte, K. Stach, M. Thiel, F. Schmidt, and H. Stork, German Offen., 2,149,070 (1973); Chem. Abstr., 79, 18434k (1973).

17. P. L. Creger, G. W. Moersch, and W. A. Neuklis, Proc. R. Soc. Med., 69, 3 (1976).

18. G. D. Diana, U. J. Salvador, E. S. Zalay, P. M. Carabateas, G. L. Williams, J. C. Collins, and F. Pancic, J. Med. Chem., 20, 757 (1977).

19. R. A. Maxwell and E. Walton, German Offen., 2,030,693 (1971); Chem. Abstr., 74, 87660q (1971).

20. R. Y. Mauvernay, N. Busch, J. Simond, A. Monteil, and J. Moleyre, German Offen., 2,310,918 (1973); Chem. Abstr., 79, 136777x (1973).

21. B. B. Molloy and M. I. Steinberg, Eur. Pat. Appl., 2,604 (1979).

22. R. A. Maxwell and F. C. Copp, German Offen., 2,030,692 (1971); Chem. Abstr., 74, 76156d (1971).

23. M. Carissimi, F. Ravenna, and G. Picciola, German Offen., 2,521,113 (1976); Chem. Abstr., 34, 164388t (1976).

24. P. J. Murphy, T. L. Williams, J. K. Smallwood, G. Bellamy, and B. B. Molloy, Life Sci., 23, 301 (1978).

25. R. W. Fleming, German Offen., 2,806,654 (1978); Chem. Abstr., 89, 197346j (1978).

26. B. Carnmalm, T. De Paulis, T. Hogberg, L. Johansson, M.-L. Persson, S.-O. Thorburg, and B. Ulff, Acta Chem. Scand. B, 36, 91 (1982); J.-E. Backvall, R. E. Nordberg, J.-E. Nystrom, T. Hogberg, and B. Ulff, J. Org. Chem., 46, 3479 (1981).

27. S. Ohki, N. Ozawa, Y. Yabe, and H. Matsuda, Chem. Pharm. Bull, 24, 1362 (1976).

28. D. J. Collins, J. J. Hobbs, and C. W. Emmens, J. Med. Chem., 14, 952 (1971).

29. D. W. Robertson and J. A. Katzenellenbogen, J. Org. Chem., 47, 2386 (1982).

4 Monocyclic Aromatic Agents

The pharmacological response elicited by monocyclic aromatic agents is a function of the number and spatial arrangement of the functional groups attached to the aromatic ring; this is true of a great many drugs.

1. ANILINE DERIVATIVES

Many local anesthetics have a selective depressant action on heart muscle when given systemically. This is useful in treatment of cardiac arrhythmias, and a lidocaine-like drug with this kind of action is <u>tocainide</u> (<u>2</u>).[1]

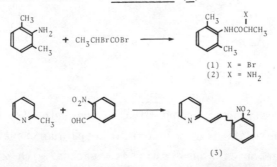

(1) X = Br
(2) X = NH_2

(3)

Part of the reason for ortho substitution in such compounds is to decrease metabolic transformation by enzymic

55

amide cleavage. Encainide (5) is another embodiment of this
concept.   Its published synthesis involves acetic anhydride-
catalyzed condensation of α-picoline with 2-nitrobenzaldehyde
to give 3. N-Methylation followed by catalytic reduction gives
piperidine 4.   The synthesis concludes by acylation with
p-methoxybenzoyl chloride to give antiarrhythmic encainide
(5).[2]

When the side chain involves an unsymmetrical urea moiety,
muscle relaxant activity is often seen. One such agent, lidamidine (6) exerts its activity as an antiperistaltic agent.
Its synthesis involves the straightforward reaction of 2,6-di-
methylphenylisocyanate and N-methylguanidine.[3]

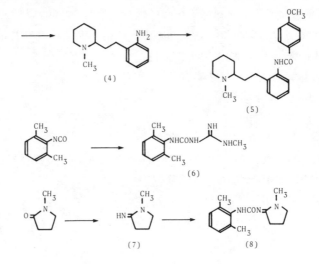

A cyclized version, xilobam (8), is synthesized from
N-methylpyrrolidone by conversion to the imine (7) by sequential reaction with triethyloxonium tetrafluoroborate and then
anhydrous ammonia. When this is reacted with 2,6-dimethyl-
phenylisocyanate, the centrally acting muscle relaxant xilobam
(8) is formed.[4]

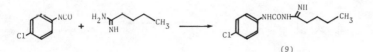

(9)

A number of muscle relaxants are useful anthelmintic agents. They cause the parasites to relax their attachment to the gut wall so that they can be eliminated. One such agent is carbantel (underline{9}). Its synthesis follows the classic pattern of reaction of 4-chlorophenylisocyanate with n-amylamidine.[5]

To prepare another such analogue, N-methylation of N,N-dicarbomethoxythiourea gives underline{10}, which itself reacts with complex aniline analogue underline{11} to give the veterinary anthelmintic agent felsantel (underline{12}).[6]

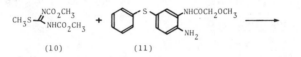

(10)   (11)

A simple aniline derivative acts as a prostatic antiandrogen. Its synthesis involves simple acylation of disubstituted aniline underline{13} with isobutyryl chloride to give flutamide (underline{14}).[7]

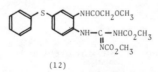

(12)

A phenylguanidine analogue is readily prepared by first reacting 2-imino-N-methylpyrrolidine with phenylisothiocyanate to give synthon underline{15}. This is next S-methylated with methyl iodide to give underline{16} which itself, on reaction with pyrrolidine, is

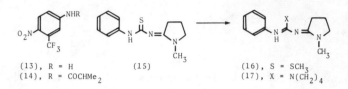

(13), R = H   (15)   (16), S = SCH_3
(14), R = $COCHMe_2$ (17), X = $N(CH_2)_4$

converted to the antidiabetic agent <u>pirogliride</u> (<u>17</u>).[8]

Finally, in demonstration of the pharmacological versatility of this chemical subclass, <u>ethyl lodoxamide</u> (<u>20</u>) shows
antiallergic properties. It shows a biological relationship
with <u>disodium chromoglycate</u> by inhibiting the release of
medi-ators of the allergic response initiated by allergens. It
can be synthesized by chemical reduction of dinitrobenzene
analogue <u>18</u> to the <u>m</u>-diamino analogue <u>19</u>. This, then, is
acylated with ethyl oxalyl chloride to complete the synthesis
of <u>ethyl lodoxamide</u> (<u>20</u>).[9]

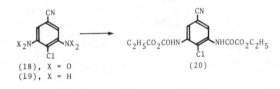

(18), X = O
(19), X = H   (20)

2. BENZOIC ACID DERIVATIVES

It has been documented in an earlier volume that appropriately
substituted molecules with two strongly electron withdrawing
substituents meta to one another in a benzene ring often
possess diuretic properties and, even though the prototypes
usually have two substituted sulfonamide moieties so disposed,
other groups can replace at least one of them. An example of
this is <u>piretanide</u> (<u>24</u>), where one such group is a carboxyl

moiety.[10]   The published synthesis starts with highly substituted benzoate 21[10] which is reduced with a Raney nickel
catalyst and converted to succinimide 23 by reaction with
succinic anhydride.

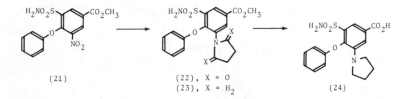

Reduction to the corresponding N-substituted pyrrolidine (23)
takes place with sodium borohydride/boron trifluoride. Saponification completes the synthesis of the diuretic agent piretanide (24).[11]

    Because of resonance stabilization of the anion, a tet-
razolyl moiety is often employed successfully as a bioisosteric
replacement for a carboxy group. An example in this subclass
is provided by azosemide (27). Benzonitrile analogue 25 is
prepared by phosphorus oxychloride dehydration of the corresponding benzamide. Next, a nucleophilic aromatic displacement
reaction of the fluorine atom leads to 26. The synthesis concludes with the 1,3-dipolar addition of azide to the nitrile
function to produce the diuretic azosemide (27).[12]

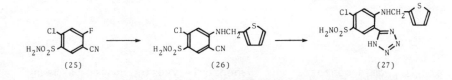

    Reversal of the amide moiety of local anesthetics is

consistent with retention of activity. So too with the derived antiarrhythmic agents. Flecainide (30) is such a substance. It is synthesized from 2,5-dihydroxybenzoic acid by base-mediated etherification with 2,2,2-trifluoroethanol. If done carefully, ester 28 results. Amide ester exchange with the appropriate pyridine amine analogues leads to 29. Catalytic reduction of the more electron-deficient aromatic ring results in the formation of flecainide (30).[13]

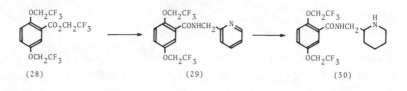

A lipid lowering agent of potential value in hyperchole-sterolemia is cetaben (31). It is synthesized facilely by monoalkylation of ethyl p-aminobenzoate with hexadecyl bromide and then saponification.[14]

Benzamide 33, known as bentiromide, is a chymotrypsin substrate of value as a diagnostic acid for assessment of pancreatic function. It is synthesized by amide formation between

ethyl p-aminobenzoate and N-benzoyl-tyrosine using N-methyl-morpholine and ethyl chlorocarbonate for activation. The resulting L-amide (32) is selectively hydrolyzed by sequential use of dimsyl sodium and dilute acid to give bentiromide (33).[15]

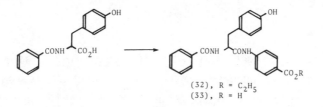

(32), R = C_2H_5
(33), R = H

3. BENZENESULFONIC ACID DERIVATIVES

As has been discussed previously, substituted p-alkylbenzene-sulfonylureas often possess the property of releasing bound insulin, thus sparing the requirement for insulin injections in adult-onset diabetes.[16] A pyrimidine moiety, interestingly, can serve as a surrogate for the urea function.

Gliflumide (37), one such agent, is synthesized from 4-isobutyl-2-chloro-pyrimidine (34) by nucleophilic displacement using p-sulfonamidobenzeneacetic acid (35) to give sulfonamide 36. Reaction, via the corresponding acid chloride, with S-1-amino-1-(2-methoxy-5-fluorophenyl)ethane completes the synthesis of the antidiabetic agent gliflumide (37).[17]

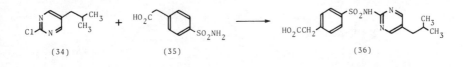

(34)   (35)   (36)

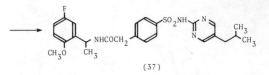

(37)

A related agent, <u>glicetanile</u> <u>sodium</u> (<u>42</u>), is made by a variant of this process. Methyl phenylacetate is reacted with chlorosulfonic acid to give <u>38</u>, which itself readily reacts with aminopyrimidine derivative <u>39</u> to give sulfonamide <u>40</u>. Saponification to acid <u>41</u> is followed by conversion to the acid chloride and amide formation with 5-chloro-2-methoxyaniline to complete the synthesis of the hypoglycemic agent <u>glicetanile</u> (<u>42</u>).[18]

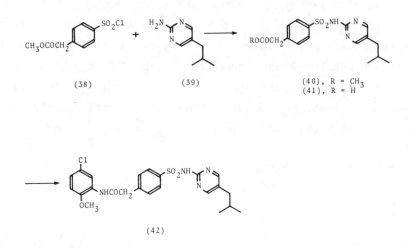

(38)   (39)   (40), R = CH$_3$
                                                (41), R = H

(42)

Perhaps surprisingly, the p-methyl benzenesulfonylurea analogue called <u>tosifen</u> (<u>45</u>), which is structurally rather close to the oral hypoglycemic agents, is an antianginal agent instead. Its synthesis involves ester-amide displacement of carbamate 43 with S-2-aminophenylpropane (44) to give 45.

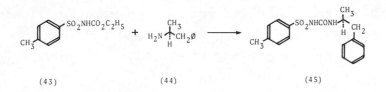

(43)   (44)   (45)

Several obvious variants exist. Tolbutamide, the prototypic drug, has some antiarrhythmic activity by an unknown mechanism. This side effect has become the principal action with tosifen, which itself does not in turn significantly lower blood sugar.[19]

REFERENCES

1. R. N. Boys, B. R. Duce, E. R. Smith, and E. W. Byrnes, German Offen. DE2,235,745 (1973); Chem. Abstr., 78, 140411v (1973).

2. H. C. Ferguson and W. D. Kendrick, J. Med. Chem., 16, 1015 (1973).

3. G. H. Douglas, J. Diamond, W. L. Studt, G. N. Mir, R. L. Alioto, K. Anyang, B. J. Burns, J. Cias, P. R. Darkes, S. A. Dodson, S. O'Connor, N. J. Santora, C. T. Tsuei, J. J. Zulipsky, and H. K. Zimmerman, Arzneim. Forsch., 28, 1435 (1978).

4. C. R. Rasmussen, J. F. Gardocki, J. N. Plampin, J. N. Twardzik, B. E. Reynolds, A. J. Molinari, N. Schwartz, W. W. Bennetts, B. E. Price, and J. Marakowski, J. Med. Chem., 21, 10 (1978).

5. G. D. Diana, French Patent FR2,003,438 (1969); Chem. Abstr., 72, 78, 7352 (1970).

6. H. Koelling, H. Thomas, A. Widdig, and H. Wollwever, German Offen. DE2,423,679 (1975); Chem. Abstr., 84, 73949k (1976).

7. J. W. Baker, G. L. Bachman, I. Schumacher, D. P. Roman, and A. L. Tharp, J. Med. Chem., 10, 93 (1967).

8. C. R. Rasmussen, German Offen. DE2,711,757 (1977); Chem. Abstr., 88, 37603s (1978).

9. J. B. Wright, C. M. Hall, and H. G. Johnson, J. Med. Chem., 21, 930 (1978).

10. D. Lednicer and L. A. Mitscher, The Organic Chemistry of Drug Synthesis, Vol. 2, Wiley, New York, 1980, p. 87.

11. W. Merkel, J. Med. Chem., 11, 399 (1976).

12. A. Popelak, A. Lerch, K. Stach, E. Roesch, and K. Hardebeck, German Offen. DE1,815,922 (1970); Chem. Abstr., 73, 45519z (1970).

13. E. H. Banitt, W. R. Bronn, W. E. Coyne, and J. R. Schmid, J. Med. Chem., 20, 821 (1977).

14. J. D. Albright, S. A. Schaffer, and R. G. Shepherd, J. Pharm. Sci., 68, 936 (1979).

15. P. L. DeBenneville and N. H. Greenberger, German Offen. DE2,156,835 (1972); Chem. Abstr., 77, 114888r (1972).

16. D. Lednicer and L. A. Mitscher, The Organic Chemistry of Drug Synthesis, Vol. 1, Wiley, New York, 1977, p. 136.

17. C. Rufer, J. Med. Chem., 17, 708 (1974).

18. K. Gutsche, E. Schroeder, C. Rufer, O. Loge, and F. Bahlmann, Arzneim. Forsch., 24, 1028 (1974).

19. L. Zitowitz, L. A. Walter, and A. J. Wohl, German Offen. DE2,042,230 (1971); Chem. Abstr., 75, 5532h (1971).

5 Polycyclic Aromatic Compounds

It will have been noted that important structural moieties are sometimes associated with characteristic biological responses (prostanoids, phenylethanolamines, for example). Just as often, however, such structural features show commonality only in the mind of the organic chemist. As will be readily evident from the very diverse biological activities displayed by drugs built on polycyclic aromatic nucleii, this classification is chemical rather than pharmacological. The nucleus does, however, sometimes contribute to activity by providing a means by which pharmacophoric groups can be located in their required spatial orientation; sometimes too, particularly in the case of the monofunctional compounds, the polycyclic aromatic moiety probably contributes to the partition coefficient so as to lead to efficient transport of the drug to the site of action.

1. INDANONES

A rather simple derivative of 1-indanone itself has been reported to possess analgesic activity. This is particularly noteworthy in that this agent, drindene (3),

departs markedly from the structural pattern of either
centrally acting or peripheral analgesics. Condensation of
1-indanone (1) with ethyl chloroformate in the presence of
alkoxide gives the corresponding hydroxymethylene derivative
2.   Reaction with ammonium acetate leads to the corresponding enamine 3, probably by addition of ammonium ion to
the terminus of the enone followed by elimination of
hydroxide.[1]

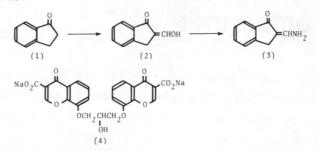

The discovery of disodium cromoglycate (4) afforded for
the first time an agent that was active against allergies by
opposing one of the very first events in the allergic
reaction; that is, the release of the various substances
(mediators) that cause the characteristic symptomology of an
allergic attack. The fact that this agent is active only by
the inhalation route led to an extensive search for a
compound that would show the same activity when administered
orally. The various candidates have as a rule been built
around some flat polycyclic nucleus and have contained an
acidic proton (carboxylate, tetrazole, etc.). One of the
simplest of these is built on an indane nucleus. Base
catalyzed condensation of phthalic ester 5 with ethyl
acetate affords indanedione 6 (shown in the enol form).
Nitration by means of fuming nitric acid leads the mediator

release inhibitor nivimedone (7).[2] The triply activated
proton shows acidity in the range of carboxylic acids.

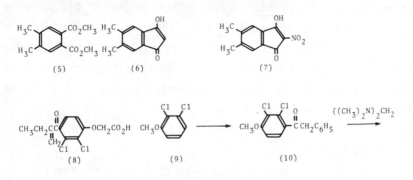

(5)   (6)   (7)

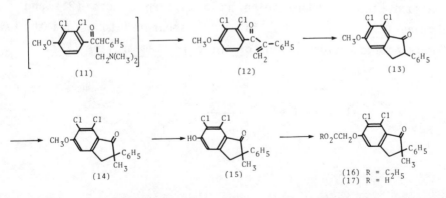

(8)   (9)   (10)

    Further investigation on the chemistry of the very
potent diuretic drug ethacrinic acid (8) led to a compound
that retained the high potency of the parent with reduced
propensity for causing side effects, such as loss of body
potassium and retention of uric acid. Friedel-Crafts
acylation of dichloroanisole 9 with phenylacetyl chloride
gives ketone 10. This is then reacted in a variant of the
Mannich reaction which involves the aminal from dimethyl-

amine and formaldehyde. The reaction may be rationalized as leading initially to the adduct 11; loss of dimethylamine leads to the enone 12. Cyclization by means of sulfuric acid affords the indanone (13). This last is in turn alkylated on carbon (14) and O-demethylated under acidic conditions. The phenol (15) thus obtained is then alkylated on oxygen by means of ethyl bromoacetate. Saponification of the ester affords indacrinone (17)[3].

2. NAPHTHALENES

As noted earlier, most classical antidepressant agents consist of propylamine derivatives of tricyclic aromatic compounds. The antidepressant molecule tametraline[4] is thus notable in that it is built on a bicyclic nucleus that directly carries the amine substituent. Reaction of 4-phenyl-1-tetralone (18) (obtainable by Friedel-Crafts cyclization of 4,4-diphenylbutyric acid) with methylamine in the presence of titanium chloride gives the corresponding Schiff base. Reduction by means of sodium borohydride affords the secondary amine as a mixture of cis (21) and trans (20) isomers. The latter is separated to afford the more active antidepressant of the pair, tametraline (20).

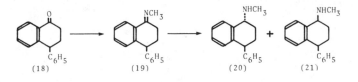

(18)   (19)   (20)   (21)

   Topical fungal infections usually involve the lipid-
like dermal and subdermal tissues. Drugs with increased
lipophilicity would thus be expected to show enhanced
antifungal activity by reason of preferential distribution
to the lipid-rich site of action. A modification of the
antifungal agent tolnaftate (29), which increases its
lipophilicity, affords tolciclate (28).   One approach to
construction of the required bridged tetrahydronaphthol (25)
involves Diels-Alder condensation of a benzyne. Thus reaction of dihalo anisole 22 with magnesium in the presence
of cyclopentadiene leads directly to the adduct 24. It is
likely that 22 initially forms a Grignard-like reagent at
the iodo group; this then collapses to magnesium halide and
benzyne 23; 1,4 addition to cyclopentadiene leads to the
observed product. Preparation of the requisite phenol 26 is
completed by catalytic hydrogenation (25) followed by O-demethylation[5].   Reaction of the sodium salt of the phenol
with thiophosgene leads to intermediate 27; condensation of
N-methyl-m-toluidine gives tolciclate (28).[6]

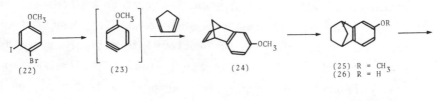

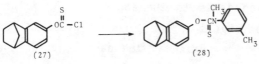
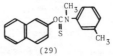

Research carried on in several laboratories in the mid-1960s indicated that triarylethylenes that carry an ethoxyethylamine substituent on one of the rings show very promising antifertility activity. It was quickly found that such agents owe their activity in the particular test system used to their ability to antagonize the effects of endogenous estrogens. One of the more potent agents synthesized in this period was nafoxidine (30). This agent's antifertility activity turned out to be restricted to rodents due to a peculiarity of the reproductive endocrinology of this species. Further clinical testing of compounds in this class revealed that certain estrogen antagonists were remarkably effective in the treatment of breast tumors, particularly those that can be demonstrated to be estrogen dependent. One such agent, tamoxifen, is currently used clinically for that indication.

More recent work in this series demonstrated that a carbonyl group can be interposed between the side-chain-carrying aromatic ring and the ethylene function with full retention of activity. Claisen condensation of benzoate 31 with 2-tetralone affords the β-diketone 32. Reaction of this with p-anisylmagnesium bromide interestingly proceeds preferentially at the ring carbonyl atom. (The thermodynamically favored enol carries the carbonyl at that position.) Spontaneous dehydration leads to the enone 33. The methoxy group on the ring substituted by the carbonyl group is rendered more reactive by the ketone at the para position thus demethylation with an equivalent of sodium ethylthiol leads to phenol 34. Alkylation with 2-chloroethylpyrrolidine affords the antiestrogenic agent trioxifene (35).[7]

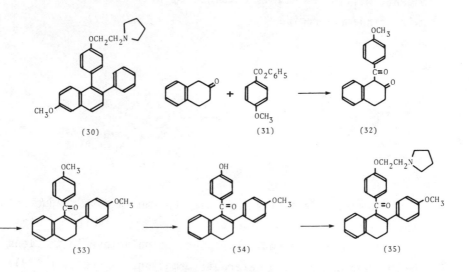

(30)   (31)   (32)

(33)   (34)   (35)

Yet another compound that exhibits antidepressant properties, that does not fit the classical mold, is a rather simple substituted amidine. Reaction of amide 36 with triethyloxonium fluoroborate (Meerwein reagent) affords the corresponding imino ether 37. Exposure of this intermediate to methylamine leads to napactidine (38).[8]

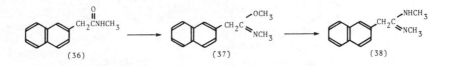

(36)   (37)   (38)

Antifungal activity has been described for an equally straightforward derivative of 1-naphthylmethylamine. This agent, naftidine (40) is obtained by alkylation of amine 39 with cinnamyl bromide.[9]

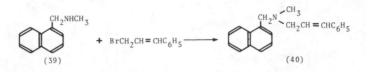

(39)   + BrCH₂CH=CHC₆H₅ ⟶   (40)

Excessive activity of the enzyme aldose reductase sometimes accompanies diabetes. The net result is often accumulation of reduced sugars such as galactose in the lens of the eye and ensuing cataract formation. Alrestatin (43), an aldose reductase inhibitor, is one of the first agents found that holds promise of preventing diabetes-induced cataracts. The compound, actually used as its sodium salt, is prepared in straightforward manner by imide formation between 1,8-naphthalic anhydride (41) and glycine.[10]

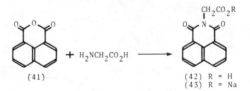

(41)   + H₂NCH₂CO₂H ⟶   (42) R = H
                                         (43) R = Na

3. TRICYCLIC COMPOUNDS: ANTHRACENE, PHENANTHRENE AND DIBENZOCYCLOHEPTENE

The discovery of the activity of the phenothiazines such as chlorpromazine (44) against schizophrenia pointed the way to drug therapy of diseases of the mind. The intensive research on the chemistry and pharmacology of those heterocycles is detailed at some length in the first Volume of this series. Those earlier investigations, it should be noted, centered mainly on modification of the side chain and substitution of the aromatic ring. Subsequent research revealed that considerable latitude exists as to the structural requirements of the central ring. For example, clomacran (45) shows the same antipsychotic activity as its phenothiazine counterpart (44). It is thus interesting to note that the fully carbocyclic analogue fluotracen (54) also exhibits CNS activity. This particular agent in fact shows a combination of antipsychotic and antidepressant activity. (In connection with the last it may be of some relevance that the compound may be viewed as a ring-contracted analogue of the tricyclic antidepressants.)

Reaction of substituted nitrile 46 with phenylmagnesium bromide gives, after hydrolysis, the benzophenone 47. Reaction of the ketone with the ylide from trimethyl-sulfoxonium iodide leads to the epoxide 48. Reductive ring opening of the oxirane by means of phosphorus and hydriodic acid completes conversion of the carbonyl to the homologous methyl group (49). Replacement of bromine by a nitrile group is accomplished by treatment of 49 with cuprous cyanide. Reaction of the product with the Grignard reagent from 3-methoxybromopropane affords the imine 51, which now contains all the required carbon atoms. Treatment of this intermediate with hybrobromic acid achieves both Friedel-Crafts-like ring closure and conversion of the terminal methoxy group to a bromide (52). The latter transformation may proceed either by direct SN_2 displacement of the protonated methoxy group by bromide or by prior cleavage of the ether to an alcohol followed by the more conventional transformation. Displacement of the terminal bromine by dimethylamine completes construction of the side chain (53). Catalytic reduction proceeds in the usual fashion to give the 9,10-dihydro derivative, fluotracen.[11] (Though not specifically stated, the method of synthesis would suggest that these groups bear a cis relationship.)

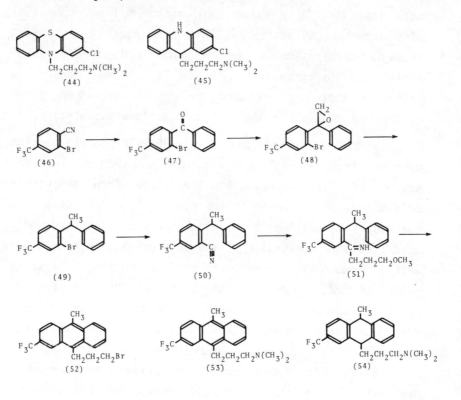

There is much evidence to suggest that one of mankind's dreaded afflictions, cancer, is not one but a loosely related series of diseases. This diversity has acted as a significant bar to the elucidation of the mechanisms underlying the uncontrolled cell division that characterizes tumor growth. Though some progress has been made toward rational design of antitumor agents, a significant portion of the drug discovery process still relies on random screening. It is thus that one of the screens sponsored by the National Cancer Institute (US) discovered the antitumor activity of a deep blue compound which had been originally synthesized as a dye for use in ball point pen ink. Preparation of this compound, <u>ametantrone</u> (<u>58</u>) starts by reaction of leucoquinizarin (<u>55</u>) with diamine <u>56</u> to give the bisimine <u>57</u>. Air oxidation of the intermediate restores the anthraquinone oxidation state. There is thus obtained <u>ametantrone</u> (<u>58</u>)[12]. A similar sequence starting with the leuco base of tetrahydroxyanthraquinone <u>59</u> affords the very potent antitumor agent <u>mitoxantrone</u> (<u>61</u>).[13]

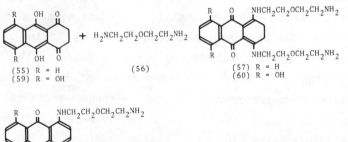

(55) R = H
(59) R = OH

(56)

(57) R = H
(60) R = OH

(58) R = H
(61) R = OH

Large-scale treatment of a host of lower organisms with biocides seems to lead almost inevitably to strains of that organism that become resistant to the effects of that agent. We thus have bacteria that no longer succumb to given antibiotics, and insects that seemingly thrive on formerly lethal insecticides. The evolution of strains of plasmodia resistant to standard antimalarial agents, coupled with the US involvement in Vietnam, a hotbed for malaria, led to a renewed search for novel antimalarial agents. Halofantrine (70) is representative of the latest generation of these compounds. The preparation of this phenanthrene starts with the aldol condensation of substituted benzaldehyde 62 with phenylacetic acid derivative 63 to give the cinnamic acid 64. Chemical reduction of the nitro group leads to aniline 65. This is then cyclized to the phenanthrene by the classical Pschorr synthesis (nitrous acid followed by strong acid). Though many methods have been proposed for direct reduction of carboxylic acids to aldehydes, these have usually been found less than satisfactory in practice. A more satisfactory method of achieving the transformation consists in reducing the acid to the carbinol (67) and then oxidizing that back to the aldehyde (68); the present sequence employs lead tetraacetate for the last step. Reformatski condensation of 68 with N,N-di-(n-butyl)bromoacetamide and zinc affords amidoalcohol 69. This is reduced to the amino alcohol by means of diborane to give halofantrine (70).[14]

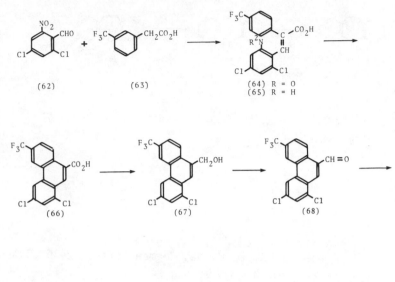

(62)   (63)   (64) R = O
                                          (65) R = H

(66)   (67)   (68)

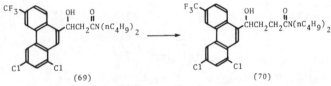

(69)   (70)

An analogue of _amitriptyline_ which contains an additional double bond in the central seven membered ring shows much the same activity as the prototype. Treatment of dibenzocycloheptanone __71__ with N-bromosuccinimide followed by triethylamine serves to introduce the additional double bond by the bromination-dehydrohalogenation sequence. Reaction of the carbonyl group with the Grignard reagent from 3-chloropropyl-N,N-dimethylamine serves to introduce the side chain (__73__). Acid catalyzed dehydration affords the antidepressant compound _cyclobenzaprine_ (__74__).[15]

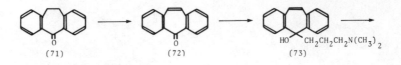

(71)   (72)   (73)

CHCH$_2$CH$_2$N(CH$_3$)$_2$
(74)

REFERENCES

1. P. D. Hammen and G. M. Milne, German Offen., 2,360,096;
    Chem. Abstr., 81, 105093 (1974).

2. D. R. Buckle, N. J. Morgan, J. W. Ross, H. Smith, and B.
    A. Spicer, J. Med. Chem., 16, 1334 (1973).

3. S. J. DeSolm, D. W. Woltersdorf, E. J. Cragoe, L. S.
    Waton and G. M. Fanelli, J. Med. Chem. 21, 437 (1978).

4. S. Reinhard, J. Org. Chem., 40, 1216 (1975).

5. H. Tanida, R. Muneyuki and T. Tsuji, Bull. Chem. Soc.
    Jap., 37, 40 (1964).

6. P. Melloni, M. Rafaela, V. Vecchietti, W. Logemann,
    S. Castellino, G. Monti, and I, DeCarneri, Eur. J. Med.
    Chem., 9, 26 (1964).

7. C. D. Jones, T. Suarez, E. H. Massey, L. J. Black and F.
    C. Tinsley, J. Med. Chem. 22, 962 (1979).

8. J. R. McCarthy, U.S. Patent 3,903,163; Chem. Abstr. 83,
    192933 (1975).

9. B. Daniel, German Offen, 2,716,943; Chem. Abstr. 88, 62215 (1978).

10. S. Kazimir, N. Simard-Duquesne and D. M. Dvornik, U.S. Patent 3,821,383; Chem. Abstr. 81, 176158 (1974).

11. P. N. Craig and C. L. Zirkle, French Patent, 1,523,230; Chem. Abstr. 71, 101609 (1969).

12. R. K. Y. Zeecheng and C. C. Cheng, J. Med. Chem., 21, 291 (1978).

13. K. C. Murdock, R. G. Child, P. F. Fabio, R. B. Angier, R. E. Wallace, F. E. Durr, and R. V. Citavella, J. Med. Chem., 22, 1024 (1979).

14. W. T. Colwell, V. Brown, P. Christie, J. Lange, C. Reece, Y. Yamamoto and D. W. Henry, J. Med. Chem, 15, 771 (1972).

15. S. O. Winthrop, M. A. Davis, G. S. Myers, J. G. Gavin, R. Thomas and R. Barber, J. Org. Chem., 27, 230 (1962).

6 Steroids

The steroid nucleus provides the backbone for both the hormones that regulate sexual function and reproduction and those involved in regulation of mineral and carbohydrate balance. The former comprise the estrogens, androgens, and progestins; cortisone, hydrocortisone, and aldosterone are the more important entities in the second category. Synthetic work in the steroid series, accompanied by some inspired endocrinological probes, led to many signal successes. Research on estrogens and progestins thus led to the oral contraceptives, and corresponding efforts on cortisone and its derivatives culminated in a series of clinically important antiinflammatory agents.

Few if any endogenous hormones seldom exert a single action. These compounds typically elicit a series of re-

sponses on different biological end points and organ systems.   It should thus not be surprising that both natural
and modified steroids also show more than one activity;
these ancillary activities, however, often consist of undesirable actions, and are thus considered side effects.
Volumes 1 and 2 of this series detail the enormous amount of
work devoted to the steroids inspired at least partly by the
goal to separate the desired activity from those side
effects.   When it became apparent that this goal might not
be achievable, there was a considerable diminution in the
synthetic work in the steroid series; this is well
illustrated by comparison of this section with its
counterparts in the preceding volumes.

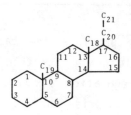

1. ESTRANES

The adventitious discovery of the antitumor action of the
nitrogen mustard poison war gases led to intensive
investigation of the mode of action of these compounds.   In
brief, it has been fairly well established that these agents
owe their effect to the presence of the highly reactive
bis(2-chloroethyl)amine group.   The cytotoxic activity of

these drugs is directly related to the ability of this group
to form an irreversible covalent bond with the genetic
material of cells, that is, with DNA. Since this alkylated
material can then no longer perform its function, replication of the cell is disrupted. The slight selectivity shown
for malignant cells by the clinically used alkylating agents
depends largely on the fact that these divide more rapidly
than those of normal tissue. There have been many attempts
to achieve better tissue selectivity by any number of other
stratagems. One of these involves linking the mustard
function to a molecule that itself shows very specific
tissue distribution. Steroids are prime candidates as such
carriers since they are well known to exhibit highly
selective organ distribution and tissue binding.
Estramustine (4) and prednimustine (64; see Section 3 below)
represent two such site directed cytotoxic agents.

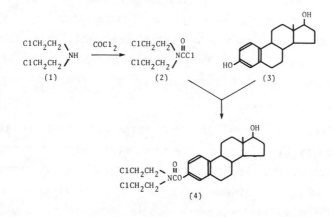

Reaction of bis(2-chloroethyl)amine with phosgene affords the corresponding carbamoyl chloride (2). Acylation of estradiol (3) with this reagent leads to estramustine (4)[1]. Though reaction with the more nucleophilic alcohol function at 17 might at first sight lead to a competing reaction, the highly hindered nature of this alcohol greatly reduces its reactivity.

Oral contraceptives almost invariably consist of a mixture of a progestational agent active by the oral route and a small amount of an estrogen. It was discovered quite early that, in contrast to the natural compounds, those lacking the 19 methyl group (19-nor) showed good oral activity; the availability of practical methods for total synthesis led to some emphasis on agents that possess an ethyl group at position 13 rather than the methyl of the natural steroids. Very widespread use of oral contraceptives led to the recognition of some side effects that were associated with the progestin component; there has thus been a trend to design drugs that could be administered in smaller quantity and a corresponding search for ever more potent progestins.

Birch reduction of the norgetrel intermediate $\underline{5}$[2] followed by hydrolysis of the enol ether gives the enone $\underline{6}$; oxidation of the alcohol at 17 leads to dione $\underline{7}$. Fermentation of that intermediate in the presence of the mold Penicillium raistricky serves to introduce a hydroxyl group at the 15 position (8). Acetal formation with neopentyl glycol affords the protected ketone which consists of a mixture of the Δ^5 and $\Delta^{5,10}$ isomers (9); hindrance at position 17 ensures selective reaction of the 3 ketone. The

hydroxyl is then converted to its mesylate (10). Treatment
with sodium acetate leads to elimination of the mesylate and
thus formation of the corresponding enone (11). Reaction of
the ketone at 17 with ethynylmagnesium bromide introduces
the requisite side chain. Removal of the ketal group by
means of aqueous oxalic acid completes the synthesis of
gestodene (13)[3].

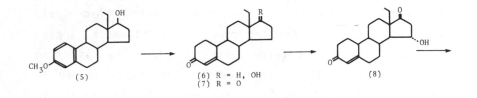

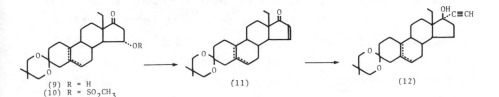

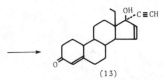

A rather more complex scheme is required for
preparation of the analogue gestrinone (27) which contains
unsaturation in rings A, B, and C. The key intermediate 24
can be obtained by Robinson annulation on dione 14 with
enone 15 to give the bicyclic intermediate 16. Successive

reduction of the double bond and the cyclopentyl carbonyl
group followed by esterification of the thus obtained
alcohol gives ketoester 17. This last can then be converted
to the enol lactone 18 by successive partial saponification
and treatment with acetic anhydride. Condensation with the
Grignard reagent from bromoketal 19 gives after hydrolysis
the tricyclic intermediate 20. (This reaction can be
rationalized as initial reaction of the organometallic with
the lactone carbonyl; the diketone formed by hydrolysis
would then cyclize under the reaction conditions.) Treatment of 20 with pyrrolidine serves to close the last ring
via its enamine (21). Hydrolysis of the first formed 3-
enamine leads to the doubly unsaturated steroid 22. Treatment with dicyanodichloroquinone (DDQ) would serve to introduce the last double bond with consequent formation of the
conjugated 5,9,10 triene (23); hydrolysis of the ester at 17
and subsequent oxidation of the alcohol would give the key
3,17 diketone 24. Treatment with ethylene glycol in the
presence of acid leads to formation of the 3 ketal.
Reaction with ethynylmagnesium bromide (26) followed by
removal of the ketal group gives gestrinone (27)[4,5].

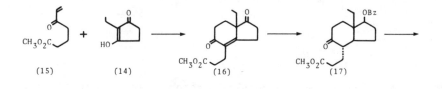

(15)   (14)   (16)   (17)

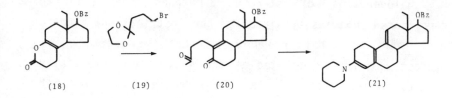

(18)   (19)   (20)   (21)

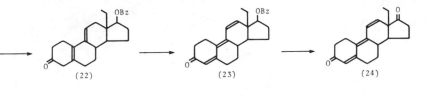

(22)   (23)   (24)

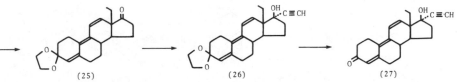

(25)   (26)   (27)

2. ANDROSTANES

Despite some early hopes, the drugs related to the androgens have found rather limited use. This has in practice been confined to replacement therapy in those cases where endogenous hormone production is deficient; some agents that show reduced hormonal effects have found some application as anabolic agents as a consequence of their ability to rectify conditions that lead to loss of tissue nitrogen. In analogy with the estrogen antagonists, an antiandrogen would seem to offer an attractive therapeutic target for treatment of diseases characterized by excess androgen stimulation (e.g., prostatic hypertrophy) and androgen dependent tumors. Attempts to design specific antagonists to androgens have met with limited success, however.

Halogenation of steroid 3-ketones can lead to complicated mixtures by virtue of the fact that the kinetic enol leads to 3 halo products, whereas the thermodynamic product is that halogenated at the 4 position. Carefully controlled reaction of the 5α-androstanolone 28 with chlorine thus leads to the 2α-chloro derivative (29). Reaction of that intermediate with O(p-nitrophenyl)-hydroxylamine affords the androgenic agent nistremine acetate (30)[6].

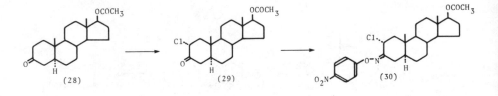

Replacement of the hydrogen at the 17 position of the prototypical androgen testosterone (31) by a propyl group interestingly affords a compound described as a topical antiandrogen. Reaction of the tetrahydropyranyl ether of dehydroepiandrosterone (32) with propylmagnesium bromide gives after removal of the protecting group the corresponding 17α-propyl derivative 33. Oppenauer oxidation of the 3-hydroxy-Δ[5] function leads to the corresponding conjugated ketone. There is thus obtained topterone (34)[7].

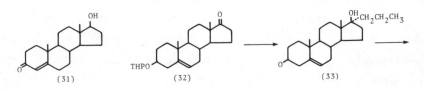

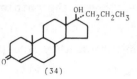

(34)

A highly modified methyltestosterone derivative also exhibits antiandrogenic activity. One synthesis of this compound involves initial alkylation of methyltestosterone (35) by means of strong base and methyl iodide to afford the 4,4-dimethyl derivative 36. Formylation with alkoxide and methyl formate leads to the 2-hydroxymethyl derivative 37. Reaction of this last with hydroxylamine leads to formation of an isoxazole ring. There is then obtained azastene (38)[8].

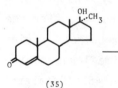

(35)

(36)

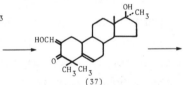

(37)

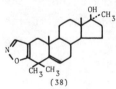

(38)

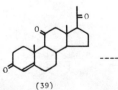

(39)

(40)

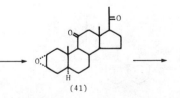

(41)

3.PREGNANES

With very few exceptions, the biological activities of synthetic steroids tend to parallel those of the naturally occurring hormones on which they are patterned. Compounds with distant pharmacological activity are, as a rule, quite rare. It is thus intriguing that inclusion of a tertiary amine at the 11 position of a pregnane leads to a compound with activity far removed from its close analogues. The agent in question, minaxalone (47), exhibits anesthetic activity. Epoxidation of progesterone derivative 40, obtainable in several steps from 11-ketoprogesterone (39)[9], gives the corresponding α-epoxide 41. Reaction of that compound with alkoxide leads to diaxial opening of the oxirane and consequent formation of the 2β-ethoxy 3α-hydroxy derivative 42. Reaction with ethylene glycol leads cleanly to selective formation of the 17 ketal (43) by reason of the highly hindered environment about the 11 carbonyl. For the same reason, formation of the oxime 44 requires forcing conditions. Chemical reduction of that oxime leads to the thermodynamically favored equatorial α-amine 45. (Catalytic reduction would have given the β-amine.) Methylation of the amine by means of formic acid and formaldehyde leads to the corresponding dimethylamino derivative (46). Removal of the ketal group completes the synthesis of minaxalone (47)[10].

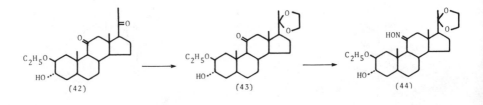

(42)   (43)   (44)

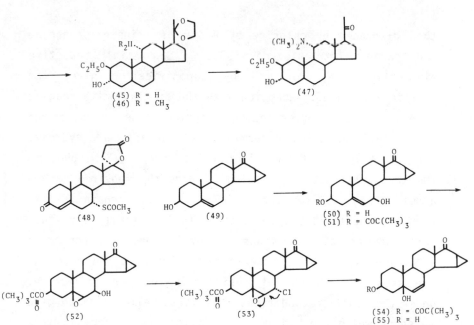

(45) R = H
(46) R = CH₃

(47)

(48)

(49)

(50) R = H
(51) R = COC(CH₃)₃

(52)

(53)

(54) R = COC(CH₃)₃
(55) R = H

Spironolactone (48) has proved a very useful diuretic
and antihypertensive agent. This drug, that owes its effect
to antagonism of the endogenous steroid hormone that regu-
lates mineral balance, aldosterone, exhibits in addition
some degree of progestational and antiandrogenic activity.
Further analogues have thus been prepared in an effort to
prepare an agent free of those side effects.

Preparation of the newest of these, spirorenone (61)[11],
starts by 7-hydroxylation of dehydroepiandrosterone derivative 49. Though this transformation has also been
accomplished by chemical means, microbiological oxidation by
Botryodiploda malorum apparently proves superior. Acylation
with pivalic anhydride proceeds selectively at the 3
hydroxyl group (51). Epoxidation by means of tertiary
butylhydroperoxide and vanadium acetylacetonate affords
exclusively the β epoxide (52). The remaining hydroxyl is

then displaced by chlorine by means of triphenylphosphine and carbon tetrachloride (53). Sequential reductive elimination (54) followed by saponification gives the allylic alcohol 55. Reaction with the Simmons-Smith reagent affords the corresponding cyclopropane, 56, the stereochemistry being determined by the adjacent hydroxyl group. Addition of the dianion from propargyl alcohol to the carbonyl group at position 17 adds the required carbon atoms for the future lactone (57). The side chain is then reduced by catalytic hydrogenation (58). Oxidation of this last intermediate by means of pyridinium chlorochromate simultaneously oxidizes the primary alcohol to an acid and the secondary alcohol at position 3 to a hydroxyketone; under the reaction conditions, the latter eliminates to give an enone while the hydroxy acid lactonizes. There is thus obtained directly the intermediate 60. Dehydrogenation by means of DDQ introduces the remaining double bond to afford spirorenone (61)[11].

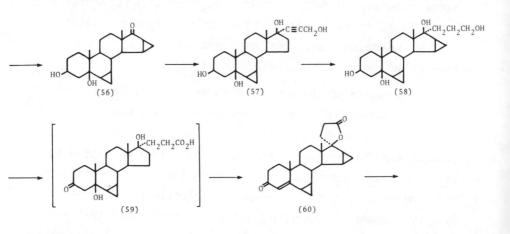

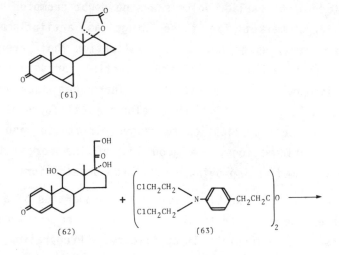

(61)

(62)   +   (63) $\longrightarrow$

As noted above, the steroid nucleus has been a favorite for the design for site directed alkylating antitumor drugs. Thus reaction of prednisolone (62) with anhydride 63 affords the 21 acylated derivative, prednimustine (64)[12].

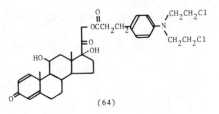

(64)

A preponderance of the work devoted to steroids, as judged from the number of compounds bearing generic names, has clearly been that in the area of corticosteroids related

to cortisone.   Much of the effort, particularly that
detailed in the earlier volumes was no doubt prompted by the
very large market for these drugs as antiinflammatory
agents. Heavy usage led to the realization that parenteral
use of these agents carried the potential for very serious
mechanism related side effects.   There has thus been a
considerable recent effort to develop topical forms of these
drugs for local application to rashes, irritation and other
surface inflammations.   A good bit of the work detailed
below is aimed at improving dermal drug penetration.

Because skin exhibits many of the properties of a lipid
membrane, dermal penetration can often be enhanced by
increasing a molecule's lipophilicity. Preparation of an
ester of an alcohol is often used for this purpose since
this stratagem simultaneously time covers a hydrophilic
group and provides a hydrophobic moiety; the ready cleavage
of this function by the ubiquitous esterase enzymes assures
availability of the parent drug molecule. Thus acylation of
the primary alcohol in <u>flucinolone</u> (<u>65</u>)[13] with propionyl
chloride affords <u>procinonide</u> (<u>66</u>)[14]; the same transform
employing cyclopropyl carbonyl chloride gives <u>ciprocinonide</u>
(<u>67</u>)[14].

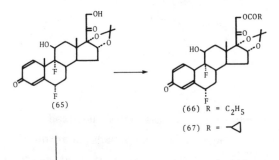

(65)

(66) R = C_2H_5

(67) R = ◁

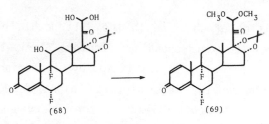

(68)   (69)

Further oxidation of the carbon at position 21 is interestingly consistent with antiinflammatory activity. Thus oxidation of <u>flucinolone</u> with copper (II) acetate affords initially the hydrated ketoaldehyde <u>68</u>; exchange with methanol gives the 21 dimethyl acetal (<u>69</u>), <u>flumoxonide</u>.[15]

Omission of the fluoro substituents at the 6 and 9 positions leads to the antiinflammatory compound agent <u>budesonide</u> (<u>71</u>). This compound is obtained by formation of the acetal of the 16,17 diol <u>70</u>[16] with butyraldehyde.[17]

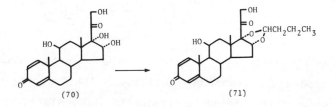

(70)   (71)

It is a hallmark of the structure activity relationships of the corticoids that the effects of structural modifications that lead to increased potency are usually additive. The fact that more than half a dozen such modifications each lead to increased potency opens ever new possibilities for combinations and permutations. <u>Meclorisone dibutyrate</u> (<u>74</u>) thus combines the known potentiating effects of the replacement of the 11-hydroxyl by chlorine as well as incorporation of a 16α-methyl

group. Addition of chlorine to the 9,11 double bond of 72
would afford the corresponding dichloro derivative, 73.
(Addition of halogen is initiated by formation of the
chloronium ion on the less hindered alpha face; diaxial
opening of the intermediate leads to the observed regio- and
stereochemistry.)   Acetylation of that intermediate with
butyric acid in the presence of trifluroacetic acid leads to
meclorisone dibutyrate (74).

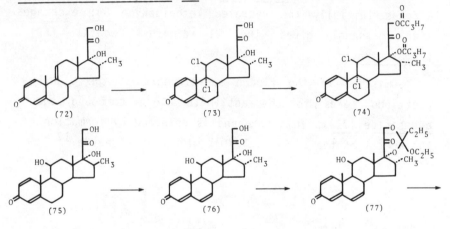

Halogenation of the 7 position also proves compatible
with good antiinflammatory activity.   Construction of this
compound, aclomethasone dipropionate (80), starts by introduction of the required unsaturation at the 6,7 position by
dehydrogenation with DDQ (76).   The highly hindered nature
of the hydroxyl at position 17 requires that a roundabout
scheme be used for formation of the corresponding ester.
Thus treatment of 76 with ethyl orthoformate affords first
the cyclic orthoformate 77. This then rearranges to the 17
ester 78 on exposure to acetic acid. Acylation of the 21
alcohol is accomplished in straightforward fashion with

propionic anhydride (79). Addition of hydrogen chloride completes the synthesis of <u>aclomethasone dipropionate</u> (80).[20] (This last reaction may in fact involve 1,6 conjugate addition of the reagent; this would then ketonize to the observed product under the acidic reaction conditions).

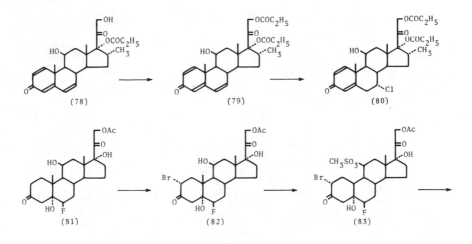

Incorporation of a vinylic bromide at the 2 position also gives a compound with good activity. Bromination of fluorohydrin 81[21] under carefully controlled conditions gives the 2α bromo derivative 82. The hydroxyl at the 11 position is then converted to its mesylate with methane-sulfonyl chloride (83). Reaction of that intermediate with acetic anhydride under forcing conditions (perchloric acid) gives the 5,17,21 triacetate 84. Treatment with sodium acetate leads to elimination of the acetate at 5 and formation of the enone 85. The presence of that function

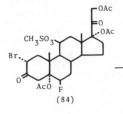

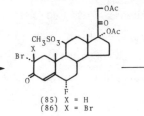

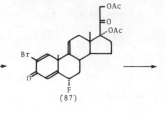

(84)

(85) X = H
(86) X = Br

(87)

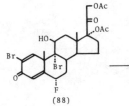

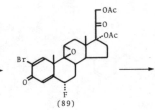

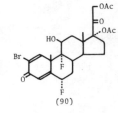

(88)

(89)

(90)

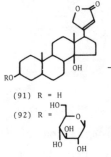

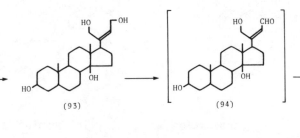

(91) R = H

(92) R =

(93)

(94)

eliminates possible future complication due to the known
facile rearrangement of halogen from the 2 to the 4
position. Thus exposure of the intermediate 85 to bromine
gives the 2,2 dibromo derivative 86. Elimination of one of
those halogens by means of lithium carbonate and lithium
bromide leads to the vinylic bromide 87; these basic
conditions achieve simultaneous elimination of the 11
mesylate and formation of the 9,11 olefin. That last

function is then used for introduction of the 9α-fluoro-11-
β-alcohol by the standard scheme. Thus exposure of the
olefin to HOBr (aqueous NBS) gives bromohydrin 88 (diaxial
opening of the initially formed α-bromonium intermediate).
Treatment with base gives the β-epoxide 89. Opening of the
oxirane with hydrogen fluoride leads finally to the
antiinflammatory agent haloprednone (90).[22]

4.MISCELLANEOUS STEROIDS

The cardiostimulant action of extracts of the leaves of the
foxglove plant (Digitalis purpurea) were recognized as early
as the eighteenth century. Careful examination of these
extracts led to the isolation of a series of so-called
cardiac glycosides which consist of hydroxylated steroids
generally substituted by an unsaturated five-membered
lactone at the 17 position and glycosidated by a series of
sugars at the 3 position. Though these drugs have proved
extremely useful in the treatment of diseases marked by
failure of the heart muscle, they are at the same time
extremely toxic. Only a very narrow margin exists between
effective and toxic doses. (It is of interest that the need
to carefully adjust blood levels of digitalis contributed to
he birth of the science of pharmacokinetics.) Though there
is a great need for a digitalis-like drug with a greater
margin of safety, there has been surprisingly little
synthetic effort devoted to this area.

It is known that the presence of the 14β-hydroxyl group
and a sugar at the 3 position are absolute requirements for
activity. A modified drug actodigin (100) demonstrates that
reversal of the lactone 17 is consistent with activity.
Reduction of digitoxigenin 91, the aglycone of digitoxin

(92) with diisobutylaluminum hydride leads to the diol 93.
Oxidation of that intermediate with triethylamine-sulfur
trioxide complex leads to the furan 95. (This transformation can be rationalized by invoking the intermediacy of an
unsaturated hydroxyaldehyde (94), followed by formation of
internal acetal 94.) The hydroxyl at the 3 position is then
protected as its chloroacetate (96). Reaction of the furan
ring with NBS followed by hydrolysis of the halide leads to
the furanone ring at 17 which is in effect the reversed
lactone from 91. The protecting group is then removed (98),
and the alcohol glycosidated with the acetylated halo sugar
99. Removal of the acetate groups by saponification affords
actodigin (101).[23]

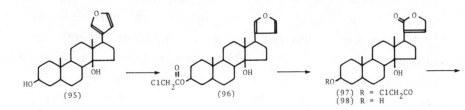

(95)   (96)   (97) R = ClCH₂CO
                                                    (98) R = H

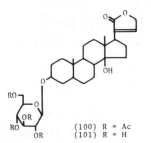

(100) R = Ac
(101) R = H

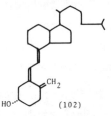

(99)

(102)

One of the triumphs of the science of nutrition is the careful investigation that linked childhood rickets with vitamin D deficiency. This work, which led to methods for treating the disease, is too familiar to need repetition. A direct consequence of these efforts was the elucidation of the pivotal role played by vitamin D in calcium metabolism, as well as the structural studies that revealed that this compound (102) is in fact a steroid derivative. The past several decades have seen the development of physical and spectroscopic methods which allow the study of ever smaller quantities of organic compounds. As a direct outgrowth of this, it has become possible to carry out very detailed studies on the metabolism of endogenous and exogenous organic compounds. Applications of such methods to vitamin D revealed that this agent in fact undergoes further hydroxylation in the body. The very high biological activity of the resulting compounds soon cast doubt on the question whether the vitamin (102) was in fact the ultimate biologically active agent. Detailed work revealed that this metabolism is in fact required for calcium regulatory action. It has been demonstrated that the mono and dihydroxy metabolites act more quickly than vitamin D and that they show much higher potency as antirachitic agents. This finding has some very practical significance since a number of disease states such as kidney failure, which are marked by calcium loss from bone, are associated with deficient vitamin D hydroxylation. The two hydroxylated metabolites, calcifediol (113) and calcitriol (145), have been introduced for clinical treatment of diseases associated with impaired vitamin D metabolism.

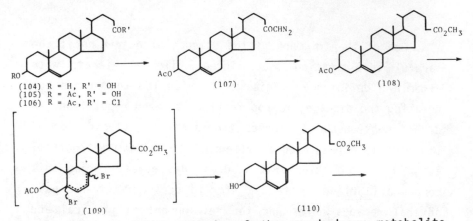

(104) R = H, R' = OH
(105) R = Ac, R' = OH
(106) R = Ac, R' = Cl

(107)

(108)

(109)

(110)

The reported synthesis of the monohydroxy metabolite (113) starts with acid 105 obtained as a by-product from oxidative cleavage of the side chain of cholesterol. This is transformed to the corresponding diazomethyl ketone 107 by reaction of the acetylated acid chloride 106 with diazomethane. Arndt-Eistert rearrangement of that intermediate affords the homologated ester 108. Allylic bromination (109) by means of dibromodimethylhydantoin followed by dehydrohalogenation with trimethyl phosphite establishes the cis diene functionality (110) required for opening of ring B. Reaction of the ester with methyl Grignard reagent completes construction of the side chain (111). Photolysis of that diene affords the product of electrocyclic ring opening (112). It now remains to isomerize the triene function. This is accomplished by thermal equilibration. There is thus obtained calcifidiol (113).[24] The low yields reported leave it open to question that this is the commercial route.

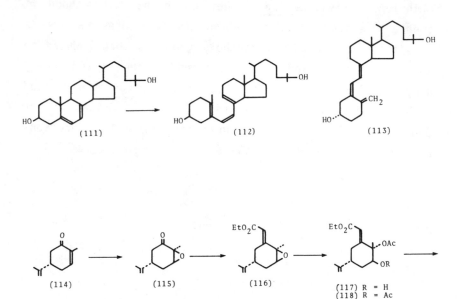

(111)   (112)   (113)

(114)   (115)   (116)   (117) R = H
                                             (118) R = Ac

A rather complex stereospecific convergent total synthesis has been reported for calcitriol (145). Construction of the A ring fragment starts with epoxidation of chiral d-carvone (114) to afford epoxide 115. Condensation with the ylide from diethyl (carboxyethyl)-phosphonate gives the corresponding ester largely as the E isomer. The epoxide is then opened with acetate (117) and acetylated to give diacetate 118. The methylene group on the side chain is then cleaved to the ketone by means of osmium tetroxide periodate reagent. Bayer-Villiger cleavage of the resulting methyl ketone (trifluoroperacetic acid) affords finally triacetate 120. This is then hydrolyzed (121) and converted to the bis trisilyl derivative 122. Dehydration under very specialized conditions gives the exo-

methylene derivative 123.   The conjugated double bond is
then isomerized by irradiation in the presence of a
sensitizer (124). The carbethoxy group is then reduced to
an alcohol (125) and this converted to the corresponding
allylic chloride (126). This reactive function is displaced
with lithium diphenylphosphide (127).   Oxidation of
phosphorus affords the A ring intermediate 128 functionalized so as to provide an ylide.

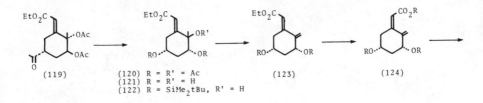

(119)

(120) R = R' = Ac
(121) R = R' = H
(122) R = SiMe$_2$tBu, R' = H

(123)

(124)

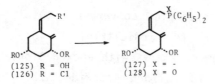

(125) R = OH
(126) R = Cl

(127) X = -
(128) X = O

The CD fragment is synthesized starting with resolved bicyclic acid 129. Sequential catalytic hydrogenation and reduction with sodium borohydride leads to the reduced hydroxy acid 130. The carboxylic acid function is then converted to the methyl ketone by treatment with methyllithium and the alcohol is converted to the mesylate. Elimination of the latter group with base leads to the conjugated olefin 133. Catalytic reduction followed by equilibration of the ketone in base leads to the saturated methyl ketone 134. Treatment of that intermediate with peracid leads to scission of the ketone by Bayer Villiger reaction to afford acetate 135. The t-butyl protecting group on the alcohol on the five membered ring is then removed selectively by means of trimethylsilyl iodide (136). Oxidation by means of pyridinium chlorochromate gives the ketone 137. That function is then reacted with ethylidine phosphorane to afford the olefin 138. Ene reaction with ethyl propiolate proceeds stereo- and regioselectively to the extended side chain of 139; note particularly that the chiral center at C_{20} has been introduced in the correct absolute configuration. Catalytic reduction leads to the saturated intermediate 140. The ester function is then reduced to the aldehyde by means of diisobutylaluminum hydride and the acetate saponified to afford 141. Condensation of the aldehyde on 141 with isopropyl phosphorane adds the last required carbon atoms (142). The tertiary hydroxyl group at the future 25 position is then introduced by means of an oxymercuration reaction (143). Oxidation of the secondary hydroxyl group completes the synthesis of the CD moiety, 144.

Condensation of the ylide from __144__ with the A ring fragment (as its TMS derivative) gives, after removal of the protecting groups, the vitamin D metabolite __cacitriol__ (__145__).[25]

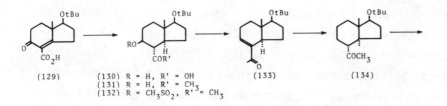

(129)   (130) R = H, R' = OH   (133)   (134)
         (131) R = H, R' = CH₃
         (132) R = CH₃SO₂, R' = CH₃

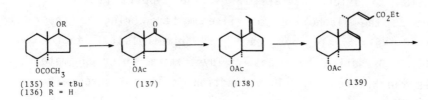

(135) R = tBu   (137)   (138)   (139)
(136) R = H

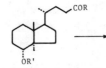 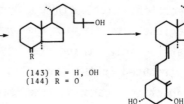

(140) R' = Ac, R = OH   (142)   (143) R = H, OH   (145)
(141) R' = H, R = H   (144) R = O

REFERENCES

1. K. B. Hogberg, H. J. Fex, I. Konyves, and H. O. J. Kneip., German Patent 1,249,862 Chem. Abstr. 68, 3118j (1968).

2. See D. Lednicer and L. A. Mitscher, The Organic Chemistry of Drug Synthesis, Vol.1, Wiley, New York, 1977, p. 168.

3. H. Hofmeister, R. Wiechert, K. Annen, H. Laurent and H. Steinbeck, German Offen. 2,546,062; Chem. Abstr. 87, 168265k (1977).

4. L. Velluz, G. Nomine, R. Bucort, and J. Mathieu, Compt. Rend., 257, 569 (1963).

5. D. Bertin and A. Pierdet, French Patent 1,503,984; Chem. Abstr. 70, 4391w (1969).

6. A. Hirsch, German Offen. 2,327,509; Chem. Abstr. 80, 83401g (1974).

7. A. L. Beyler and R. A. Ferrari, Ger. Offen., 2,633,925; Chem. Abstr., 86 161310s (1977).

8. G. O. Potts, U.S. Patent 3,966,926; Chem. Abstr. 85, 83244 (1976); note this is a use patent.

9. D. Lednicer and L. A. Mitscher, The Organic Chemistry of Drug Synthesis, Vol. 1, Wiley, New York, 1977, p. 191.

10. G. H. Phillips and G. Ewan, German Offen. 2,715,078; Chem. Abstr. 88, 38078m (1978).

11. D. Bittler, H. Hofmeister, H. Laurent, K. Nickiseh, R. Nickolson, K. Petzold and R. Wiechert, Angew. Chem. Int. Ed. Engl. 21, 696 (1982).

12. H. J. Fex, K. B. Hogberg and I. Konyves, German Offen. 2,001,305; Chem. Abstr. 73, 99119n (1970).

13. D. Lednicer and L. A. Mitscher, The Organic Chemistry of Drug Synthesis, Vol. 1, Wiley, New York, 1977, p. 202.

14. B. J. Poulson, U.S. Patent 3,934,013; Chem. Abstr. 84, 111685f (1976).

15. M. Marx and D. J. Kertesz, German Offen. 2,630,270; Chem. Abstr., 86, 140,345s (1977).

16. D. Lednicer and L. A. Mitscher, The Organic Chemistry of Drug Synthesis, Vol. 2, Wiley, New York, 1980, p. 180.

17. A. Thalen and R. Brattsand, Arzneim Forsch. 29, 1607 (1979).

18. D. Lednicer and L. A. Mitscher, The Organic Chemistry of Drug Synthesis, Vol. 1, Wiley, New York, 1977, p. 198.

19. E. Shapiro, et.al. Steroids 9, 143 (1967).

20. M. J. Green, H. J. Shue, E. L. Shapiro and M. A. Gentles, U.S. Patent 4,076,708; Chem. Abstr. 89, 110119r (1978).

21. D. Lednicer and L. A. Mitscher, The Organic Chemistry of Drug Synthesis, Vol. 1, Wiley, New York, 1977 p. 195.

22. M. Riva and L. Toscano; German Offen. 2,508,136; Chem. Abstr. 84, 31311 (1976).

23. J. M. Ferland, Can. J. Chem. 52, 1652 (1974).

24. J. A. Campbell, D. M. Squires, and J. C. Babcock, Steroids 13, 567 (1969).

25   E. G. Baggiolini, J. A. Iacobelli, B. M. Hennessy, and M. V. Uskokovic, J. Am. Chem. Soc., 104, 2945 (1982).

7 Compounds Related to Morphine

Pain is probably the immediate stimulus for more visits to the physician's office than all other complaints combined. Since pain serves as an alert to injury, it is often the first harbinger of disease; pain is thus associated with a multitude of physical ills. The fact that this sensation often persists well beyond the point where it has served its alerting function makes its alleviation a prime therapeutic target. In a very general way, the sensation of pain can be divided into two segments. The first is the immediate stimulus that sets off the chain of events; this could be a surface injury such as a burn or a cut, an inflamed internal organ, or any other disorder that causes pain receptors to be triggered. Following a rather complex series of neurochemical transmissions, the signal reaches the brain, where it is processed and finally presented as the sensation of pain. Treatment of pain follows a roughly similar duality. The pain receptors, for example, can be blocked by local anesthetics in those few cases where the insult is localized in the periphery of the body. (In practice, this is restricted to minor surgery and dentistry.) It has recently become recognized that the pain that accompanies inflammation and related conditions is actually triggered by the local synthesis of high levels of prostaglandins; compounds that inhibit this reaction will, in fact, attenuate the pain attendant to the causative inflammation. Cyclooxygenase inhibitors such as aspirin and other nonsteroid antiinflammatory agents have thus found a secure place in the treatment of the mild to moderate pain associated with elevated prostaglandin synthesis. There remains, however, a large category of pain that is not affected by interference at the receptor stage; intervention is achieved in this case at the level of the central nervous system; hence the sobriquet, central analgetics. Rather than blocking or in some way interfering with the pain signal, agents in this class change the perception of the signal. The opium alkaloid morphine (1) is the prototype central analgetic. The fact that its analgetic properties were discovered centuries ago make it clear that in this case theory came a good bit later than practice.

(1)

Though morphine is an extremely effective analgesic, it
has an associated series of side effects that limit its
legitimate use.   The most prominent among these is, of
course, its propensity to cause physical addiction.   A
significant amount of work has thus been devoted to the
synthesis of analogues with a view to modifying the
pharmacological spectrum and, in particular, avoiding
addiction potential.   As will be noted from the following
discussion (and that in the earlier volumes), this work has
led to structures that have little in common with the
prototype molecule.

1.BRIDGED POLYCYCLIC COMPOUNDS
It has been found empirically that central analgesics that
possess some degree of activity as antagonists of the
effects of morphine tend to show a reduced propensity for
causing physical addiction. Again empirically, it was noted
that this could often be achieved by replacement of the N-
methyl   group   by   allyl,   cyclopropylmethyl,   or
cyclobutylmethyl; additional nuclear modifications often
contributed to this activity.

Exposure of the opium alkaloid thebaine (2) to mild
acid leads to hydrolysis of the enol ether function followed
by migration of the double bond to yield the conjugated
enone 3. Addition of lithium diethylcuprate proceeds by 1,4
addition from the less hindered side to give the
intermediate 4. Treatment of that with cyanogen bromide
under von Braun conditions leads to the isolable
aminocyanide (5). This is then coverted to the secondary
amine (6) by treatment with aqueous base. Alkylation of

that intermediate with cyclopropylmethyl chloride affords
the analgesic codorphone (underline{7}).[1]

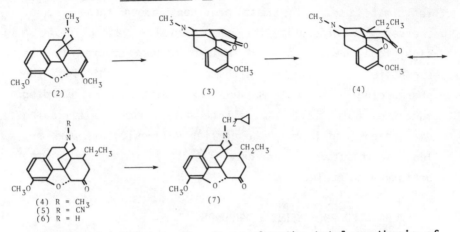

The development of schemes for the total synthesis of
the carbon skeleton of morphine revealed that the fused
furan ring was not necessary for biological activity. More
recently it has been found that substitution of a pyran ring
for the terminal alicyclic is also consistent with
biological activity. Starting material for this preparation
is ketoester underline{8}, available by one of the classical
benzomorphan syntheses.[2] Condensation with the ylide from
diethyl(carbethoxyethyl)phosphonate affords diester underline{9}. (The
course of the reaction is probably helped by the fact that
the β-ketoester can not undergo tautomerism to its enol
form.) Catalytic reduction proceeds from the less hindered
face to afford the corresponding saturated diester (underline{10}).
Reduction of the carbonyl function by means of lithium
aluminum hydride gives the glycol underline{11}; this undergoes
internal ether formation on treatment with acid to form the
pyran ring of underline{12}. Treatment with cyanogen bromide (or ethyl
chloroformate) followed by saponification of the

intermediate leads to the secondary amine (14). This is converted to the cyclopropylmethyl derivative 16 by acylation with cyclopropylcarbonyl chloride followed by reduction of the thus formed amide (15) with lithium aluminum hydride. Cleavage of the O-methyl ether with sodium ethanethiol affords proxorphan (17).

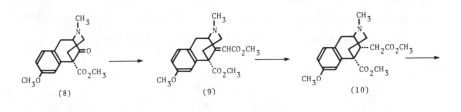

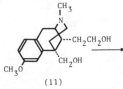

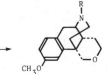

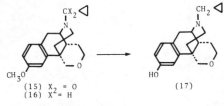

Replacement of the alicyclic ring of morphine in addition to omission of the furan ring leads to a thoroughly investigated series of analgesic compounds known as the

benzomorphans. Depending on the substitution pattern, these
agents range in activity from potent agonists to
antagonists. Reduction of the carbonyl group in oxygenated
benzomorphan 18 affords the corresponding alcohol (19).
This intermediate is then N-demethylated by means of
cyanogen bromide (20). Acylation with cyclopropylcarbonyl
chloride gives the amide 21. The alcohol is then converted
to the ether 22 by treatment with methyl iodide and base.
Treatment with lithium aluminum hydride serves to reduce the
amide function (23). Cleavage of the phenolic ether by one
of the standard schemes affords moxazocine (24).[4]

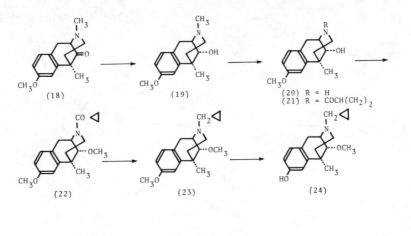

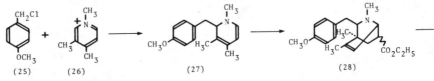

A rather unusual reaction forms the key step in the
preparation of a benzomorphan bearing a fatty side chain.
The scheme used to form the benzomorphan nucleus, which is
patterned after the Grewe synthesis originally developed for

preparing morphinans, is fairly general for this class of compounds. Preparation of tonazocine (33) starts with the condensation of the Grignard reagent from 25 with the pyridinium salt 26. (Note that reaction actually occurs at the more sterically hindered center.) In the usual synthetic route, the enamine function would next be reduced and the olefin cyclized. In the case at hand, however, the diene function of 27 is condensed with ethyl acrylate in a Diels-Alder reaction (28). Treatment with strong acid leads to cyclization of the olefin into the aromatic ring and formation of the benzomorphan nucleus (29). Acylation of the anion obtained from that intermediate by means of lithium diisopropyl amide with hexanoyl chloride gives the β-ketoester 30. Heating of that compound with formic acid leads directly to the ring-opened benzomorphan 32. This transform can be rationalized as involving proton mediated cleavage of the blocked β-ketoester to afford 31; decarbethoxylation under the strongly acidic conditions then leads to the observed product. Cleavage of the phenolic ether affords the analgesic agent tonazocine (33).[5]

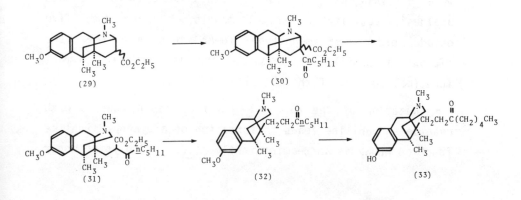

2.PIPERIDINES

Still further simplification of the structural requirements for central analgesic activity came from the serendipitous observation that the simple phenylpiperidine, meperidine (34), shows biological activity almost indistinguishable from that of morphine. Further elaboration of that molecule led to one of the most potent opioid analgesics, fentanyl (35).

(34)

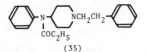

(35)

In-depth investigation of the structure activity relationships in the fentanyl meperidine series in the Janssen laboratories revealed that additional substitution of the amide nitrogen-bearing center resulted in still further enhancement of analgesic potency. Several compounds were obtained as a result of this work, which showed analgesic activity in animal models at doses some five decade orders of magnitude lower than morphine; the biological profile of these agents is, however, almost identical to that of the classical opioids.

Reaction of the carbonyl group of piperidone 36 with cyanide and aniline leads to formation of a cyanohydrin-like function known as an α-aminonitrile (37); hydrolysis under strongly acidic conditions gives the corresponding amide 38. (Although aminonitriles are somewhat labile, particularly under basic conditions, the corresponding aminoamide is a quite stable function.) The benzyl amine protecting group is then removed by catalytic hydrogenation (39). Alkylation with 2-phenethyl bromide proceeds on the more nucleophilic aliphatic amine to afford 40. Ethanolysis of the amide function leads to the corresponding ester, 41. Acylation of the remaining secondary amine function with propionyl chloride affords carfentanyl (42).[6] The same sequence starting with the corresponding 3-methylpiperidone (43)[7] affords lofentanyl (44).

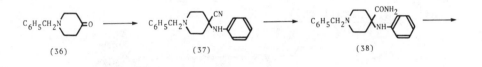

(36)   (37)   (38)

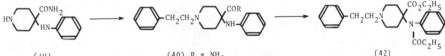

(39)   (40) R = NH₂   (42)
              (41) R = OC₂H₅

(43)

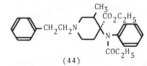

(44)

Alkylation of intermediate <u>39</u> with 2-(2-bromo-
ethyl)thiophene affords the corresponding thiophene-
containing compound <u>45</u>. Ethanolysis then leads to ethyl
ester (<u>46</u>). Reduction of the carbonyl function affords the
carbinol <u>47</u>. Alkylation of the alkoxide obtained from the
alcohol with methyl iodide gives the methyl ether <u>48</u>.
Acylation with propionyl chloride leads to the very potent
opioid analgesic <u>sulfentanyl</u> (<u>49</u>).[8] In the absence of a
specific reference, one may speculate that alkylation of
heterocycle <u>50</u> with 1-bromo-2chloroethane would give the
chloroethyl derivative <u>51</u>. Use of this for alkylation of <u>39</u>
would give the heterocycle substituted intermediate 52.   .
A similar scheme via structures <u>53</u>, <u>53a</u>, and <u>54</u> would then
afford <u>alfentanil</u> (<u>55</u>).

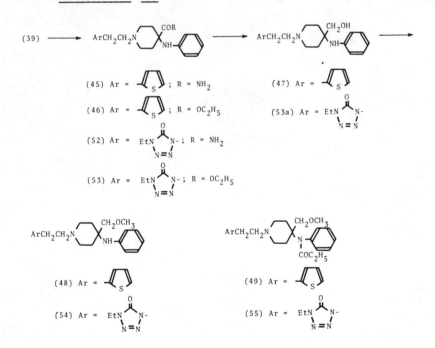

   Fusion of an alicyclic ring onto the piperidine so as
to form a perhydroisoquinoline is apparently consistent with
analgesic activity.   Synthesis of this agent, ciprefadol
(68), starts with the Michael addition of the anion from
cyclohexanone 56 onto acrylonitrile (57). Saponification of
the nitrile to the corresponding acid (58) followed by
Curtius rearrangement leads to isocyanate 59.   Acid
hydrolysis of the isocyanate leads directly to the indoline
61, no doubt by way of internal Schiff base formation from
the intermediate amine 60.   Alkylation by means of
trimethyloxonium fluoroborate affords ternary iminium salt
62.   Treatment of that reactive carbonyl-like functionality
with diazomethane gives the so-called azonia salt 63 (note
the analogy to the hypothetical oxirane involved in ring
expansion of ketones with diazomethane).   Exposure of the
aziridinium intermediate to base leads to ring opening and
consequent formation of the octahydroisoquinoline (64).
Reduction of the enamine (catalytic or borohydride) affords
the perhydroisoquinoline 65.   This compound is then
subjected to one of the N-demethylation sequences and the
resulting   secondary   amine   (66)   alkylated   with
cyclopropylmethyl bromide (67).   O-Demethylation of the
phenol ether completes the preparation of ciprefadol (68).[9]

(50)   (51)

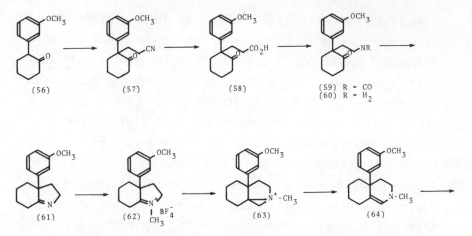

(56)   (57)   (58)   (59) R = CO
                                          (60) R = H₂

(61)   (62)   (63)   (64)

An internally bridged arylpiperidine in which the aryl
group has been moved to the 3 position interestingly retains
analgesic activity. It should be noted, however, that this
compound has been described as nonnarcotic on the basis of
its animal pharmacology. Reaction of the carbene obtained
from treatment of bromoaryl acetate 69 with ethyl acrylate
affords the cyclopropane 70. The cis stereochemistry of the
product probably represents the fact that this isomer shows
fewer nonbonding interactions than does its trans counterpart. Saponification of the ester followed by reaction of
the resulting diacid (71) with urea leads to the cyclic
imidide 72. Reduction of the carbonyl groups is achieved by
treatment with sodium aluminumbis[2(methoxy)ethoxy]hydride.
There is thus obtained bicifadine (73).[10]

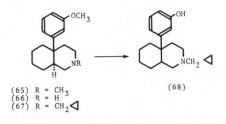

(65) R = CH₃   (68)
(66) R = H
(67) R = CH₂◁

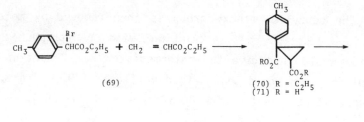

(69)

(70) R = C$_2$H$_5$
(71) R = H

(72)   (73)

3.MISCELLANEOUS COMPOUNDS

The large amount of synthetic work devoted to central analgesic agents led to the elaboration of fairly sound structure activity relationships. Until fairly recently, structural requirements for analgesic activity could be reliably covered by the so-called Beckett and Casey rule. In its most general form, this requires an aromatic ring attached to a quaternary center with basic nitrogen at a distance of about two carbon atoms from that center. (Fentanyl and its analogues were incorporated by assuming that the fully substituted anilide nitrogen is equivalent to the quaternary center.) Some quite potent analgesics that have recently been prepared do not fit this generalization very well, suggesting that it perhaps needs to be revised.

The benzazepines, verilopam (79) and anilopam (81), for example, represent significant departures from the above generalization. Construction of the former starts with the alkylation of veratrylamine (74) with the dimethyl acetal of bromoacetaldehyde to give the secondary amine 75. Cyclization under acidic conditions leads to the benzazepine

76. The benzylic methoxy group is then removed by metal-ammonia reduction.[11]   Alkylation with p-nitrophenethyl bromide would then give the intermediate 78. Reduction of the nitro group would thus afford verilopam (79). The same sequence starting with amine 80 affords anilopam (81).

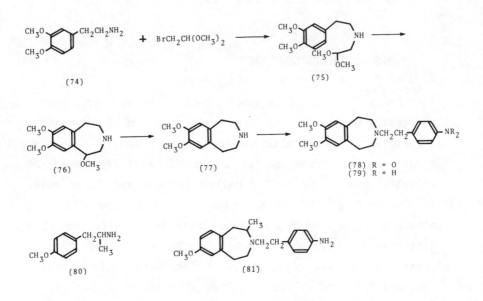

The good analgesic efficacy observed with ciramadol (87) and doxpicomine (92) show that location of the basic center directly on the quaternary benzylic center is quite consistent with activity. It is interesting to note in this connection that compound 82, in which nitrogen is similarly located, shows analgesic potency in the range of sulfentanyl that is, some five decade orders of magnitude greater than morphine.[12]

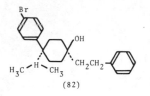

(82)

Aldol condensation of the methoxymethyl ether of m-methoxybenzaldehyde (83) with cyclohexanone affords the conjugated ketone 84. Michael addition of dimethylamine leads to the aminoketone 85. Reduction of the ketone proceeds stereospecifically to afford the cis aminoalcohol 86. Mild hydrolysis of the product gives the free phenol, ciramadol (87).[13]

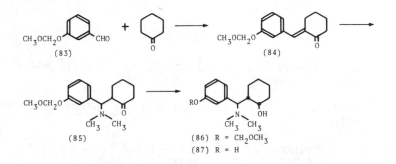

(83)   (84)

(85)   (86) R = CH$_2$OCH$_3$
            (87) R = H

In a similar vein, Knoevenagel condensation of nicotinaldehyde (88) with diethyl malonate gives the unsaturated ester 89. Michael addition of dimethylamine gives the corresponding aminoester (90). Reduction of the carbonyl groups with lithium aluminum hydride affords the glycol 91. Formation of the acetal between the diol and acetone gives the analgesic agent doxpicomine (92).[14]

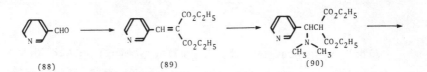

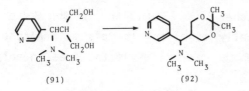

REFERENCES

1. M. P. Kotick, D. L. Leland, J. O. Polazzi, and R. N. Schut, J. Med. Chem., 23, 166 (1980).

2. See, for example, J. A. Baltrop, J. Chem. Soc., 1947 349.

3. T. A. Montzka, J. D. Matiskella and R. A. Partyka, U.S. Patent 4,246,413; Chem. Abstr. 95, 43442z (1981).

4. T. A. Montzka and J. D. Matiskella, German Offen. 2,517,220, Chem. Abstr. 84, 59832k (1976).

5. W. F. Michne, T. R. Lewis, S. J. Michalec, A. K. Pierson, and F. J. Rosenberg, J. Med. Chem., 22, 1158 (1979).

6. P. G. H. VanDaele, M. L. F. DeBruyn, J. M. Boey, S. Sanczuk, J. T. M. Agten, and P. A. J. Janssen, Arzneim. Forsch., 26 1521 (1971).

7. W. F. M. VanBever, C. J. E. Niemegeers, and P. A. J. Janssen, J. Med. Chem., 17, 1047 (1974).

8. W. F. M. VanBever, C. J. E. Niemegeers, K. H. Schellekens, and P. A. J. Janssen, Arzneim. Forsch., 26, 1548 (1976).

9. D. M. Zimmerman and W. S. Marshall, U.S. Patent 4,029,796; Chem. Abstr. 87, 102192 (1977).

10. J. W. Epstein, H. J. Brabander, W. J. Fanshawe, C. M. Hofmann, T. C. McKenzie, S. R. Safir, A. C. Usterberg, D. B. Cosulich, and F. M. Lovell, J. Med. Chem., 24, 481 (1981).

11. T. A. Davison and R. D. Griffith, German Offen 2,946,794; Chem. Abstr. 93, 186198 (198).

12. D. Lednicer and P. F. VonVoigtlander, J. Med. Chem., 22, 1157 (1979).

13. J. P. Yardley, H. Fletcher, III, and P. B. Russell, Experientia 34, 1124 (1978).

14. R. N. Booher, S. E. Smits, W. W. Turner, Jr., and A. Pohland, J. Med. Chem., 20, 885 (1977).

8 Five-Membered Heterocycles

A five-membered heterocyclic ring packs a relatively large number of polarized bonds into a relatively small molecular space. This provides a convenient framework to which to attach necessary side chains. In some cases, the framework itself is believed to be part of the pharmacophore.

1. PYRROLES AND PYRROLIDINES

In recent years increasing attention has been paid to the possibility of delaying or even reversing the memory loss that accompanies old age or the more tragic loss of human capabilities associated with premature senility - Alzheimer's disease. Progress is hampered by the difficulty of identifying suitable animal tests, and there is presently no reliable therapy.

A series of pyrrolidones shows promise of being cognition-enhancing agents. One of these, amacetam (3), is synthesized readily by ester-amide exchange between ethyl 2-oxo-1-pyrrolidineacetate (1) and N,N-diisopropylethylenediamine (2).[1]

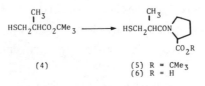

(1)   (2)   (3)

A tragic amount of morbidity and mortality is associated with high blood pressure. Many drugs operating by a wide variety of mechanisms have been employed to control this condition. Of the biochemical mechanisms employed by the body to maintain blood pressure an important one involves conversion of a kidney protein, angiotensinogen, to the pressor hormone angiotensin by a series of enzymes. The last step in the activation involves cleavage of angiotensin I to the much more active angiotensin II by the so called angiotensin-converting enzyme. It was believed that inhibitors of converting enzyme would have antihypertensive activity. Captopril (6), designed expressly for this purpose, has found exceptional clinical acceptance. One of the several syntheses of this fairly simple molecule involves amidation of t-butyl prolinate by 3-thio-2-methylpropionic acid (4) followed by acid treatment of the protected intermediate (5) to give captopril (6).[2]

$$\underset{(4)}{HSCH_2\overset{CH_3}{\underset{|}{CH}}CO_2CMe_3} \longrightarrow \underset{\substack{(5) \ R = CMe_3 \\ (6) \ R = H}}{HSCH_2\overset{CH_3}{\underset{|}{CH}}CON\diagdown_{CO_2R}}$$

A nonsteroidal antiinflammatory agent in which the benzene ring carrying the acetic acid moiety has been replaced by a pyrrole grouping is zomepirac (10). It is synthesized from

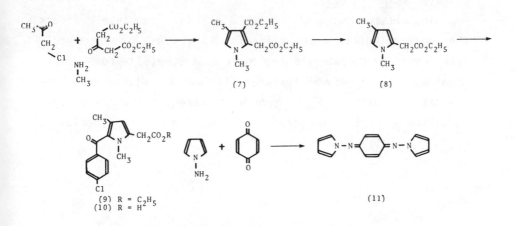

(7)   (8)

(9) R = C_2H_5
(10) R = H

(11)

diethyl acetonedicarboxylate, chloroacetone, and aqueous methylamine _via_ modification of the Hantsch pyrrole synthesis to
give key intermediate _7_. Saponification, monoesterification
and thermal decarboxylation give ester _8_. This is acylated
with _N_,_N_-dimethyl-_p_-chlorobenzamide (to give _9_) and, finally,
saponification gives zomepirac (_10_).[3]

Treatment of _p_-benzoquinone with 1-pyrrolidinylamine provides a convenient synthesis of the immunoregulator and antibacterial agent, azarole (_11_).[4]

2. FURANS

A biarylpropionic acid derivative containing a furan ring as a
prominent feature has antiinflammatory activity. The patented
synthesis involves a straightforward organometallic addition of
ethyl lithioacetate to aldehyde _12_ followed by saponification

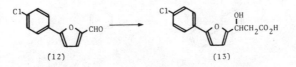

(12)   (13)

to give <u>orpanoxin</u> (<u>13</u>).[5]

Installation of a different side chain completely alters the pharmacological profile leading to a new class of muscle relaxants. The synthesis begins with copper(II)-promoted diazonium coupling between furfural (<u>14</u>) and 3,4-dichlorobenzene-diazonium chloride (<u>15</u>) to give biarylaldehyde <u>16</u>. Next, condensation with 1-aminohydantoin produces the muscle relaxant <u>clodanolene</u> (<u>17</u>).[6]

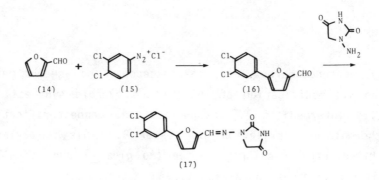

The pharmacological versatility of this general substitution strategy is further illustrated by diazonium coupling of <u>14</u> with 2-nitrobenzenediazonium chloride to produce biarylaldehyde <u>18</u>. Formation of the oxime with hydroxylamine is followed by dehydration to the nitrile. Reaction with anhydrous methanolic hydrogen chloride leads to imino ether <u>19</u> and addition-elimination of ammonia leads to the antidepressant amidine, <u>nitrafudam</u> (<u>20</u>).[7]

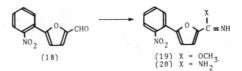

The presence of a furan ring is also compatible with

cimetidine-like antiulcer activity, despite the prominent pos-
session of an imidazole ring by histamine, the natural agonist,
which served as a structural point of departure for the histamine H_2 antagonists. The synthesis of ranitidine (25) begins
with an acid-catalyzed displacement of the primary alcoholic
linkage of 21 with cysteamine to give 22. An addition-elimination reaction with 23 (itself made from 24 by an addition-
elimination reaction) completes the synthesis of ranitidine
(25).[8]

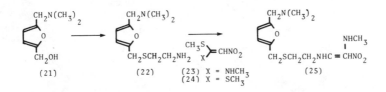

3. IMIDAZOLES

Amoebal infections, particularly of farm animals and the female
human genitalia, are at best only annoying. All too often the
problem encountered leads to difficult diarrheas. A group of
nitroimidazoles have activity against the causative organisms
and consequently have been widely synthesized.

One such agent is synthesized from 2-methylimidazole by
reaction with epichlorohydrin under acidic conditions. This
produces the antiprotozoal agent ornidazole (26).[9]

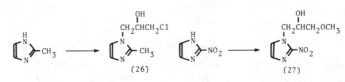

Similarly, reaction of 2-nitroimidazole with 1,2-epoxy-3-methoxypropane in the presence of potassium carbonate gives misonidazole (27).[10] This agent also has the interesting and potentially useful additional property of sensitizing hypoxic tumor cells to ionizing radiation.

Many nitroimidazoles possess antiprotozoal activity. One of these is bamnidazole (29). Synthesis involves reaction of imidazole carbonate 28 with ammonia.[11]

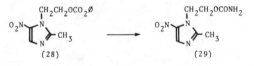

Removal of the nitro group results in an alteration of antimicrobial spectrum leading to a series of antifungal agents. For example, reaction of 2,4-dichloroacetophenone with glycerol and tosic acid leads to dioxolane 30. Under brominating conditions, sufficient carbonyl-like character exists to allow transformation of 30 to 31 and this product, after esterification, undergoes displacement to 32 with imidazole. Saponification and reaction with mesyl chloride then give 33. The synthesis of antifungal ketoconazole (34) then concludes by displacement with the phenolate derived from 4-acetylpiperazinylphenol.[12]

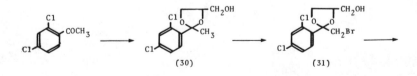

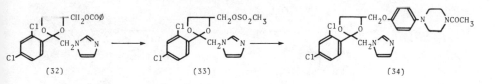

(32)   (33)   (34)

Displacement of bromine on phenacyl halide **35** with imidazole gives **36**. Reduction with sodium borohydride followed by
displacement with 2,6-dichloro-benzyl alcohol in HMPA then produces antifungal orconazole (**37**).[13]

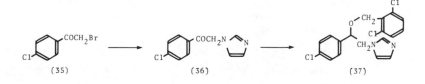

(35)   (36)   (37)

If the displacement reaction is carried out between imidazole derivative **38** and thiophene analogue **39**, the antifungal
agent tiaconazole (**40**) results.[14] A rather slight variant of
this sequence produces antifungal sulconazole (**41**).[15] Obvious
variants of the route explicated above for ketoconazole (**34**)
lead to parconazole (**42**)[16] and doconazole (**43**),[17] instead.

Insertion of a longer spacer is compatible with antifungal
activity. Reaction of epichlorohydrin with 4-chlorobenzyl-
magnesium chloride leads to substituted phenylbutane **44**. Dis-

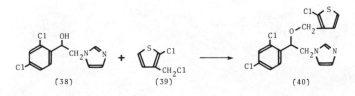

(38)   (39)   (40)

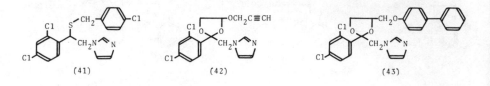

(41)   (42)   (43)

placement with sodium imidazole, conversion of the secondary
alcohol group to the chloride (thionyl chloride), and displacement with 2,6-dichlorothiophenolate concludes the synthesis of
antifungal <u>butoconazole</u> (<u>45</u>).[18]

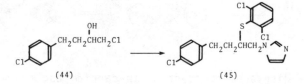

(44)   (45)

Progressive departure from the fundamental structure of
the lead agent <u>cimetidine</u> led to the antiulcer agent <u>oxmetidine</u>
(<u>47</u>). The synthesis involves <u>S</u>-methylation (CH_3I) of the 2-
thiouracil intermediate <u>46</u> and is followed by an addition-
elimination reaction with 2-(5-methyl-4-imidazolylmethylthio)
ethylamine to give <u>oxmetidine</u> (<u>47</u>).[19]

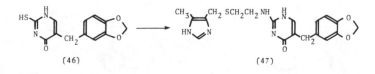

(46)   (47)

Another entry into the antiulcer sweepstakes is <u>etinfidine</u>
(<u>50</u>). It is synthesized by displacement of halide from 4-
chloromethyl-5-methylimidazole (<u>48</u>) with substituted thiol <u>49</u>.
The latter is itself made from thiourea analogue <u>51</u> by an addition-elimination reaction with cysteamine <u>52</u>.[20]

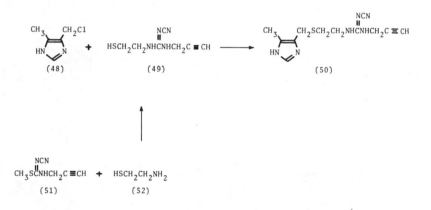

The imidazole-containing hypnotic/injectable anesthetic
agent <u>etomidate</u> (<u>58</u>) is synthesized from 1-amino-1-phenylethane
starting with triethylamine mediated displacement with chloro-
acetonitrile leading to secondary amine <u>53</u>. The <u>d</u>-enantiomer
is preferred as starting material. This is converted to the
formamide (<u>54</u>) on heating with formic acid. Next, the active
methylene group is formylated by reaction of <u>54</u> with ethyl
formate and sodium methoxide in order to give <u>55</u>. The now
superfluous <u>N</u>-formyl group is removed and the imidazole ring is
established upon reaction of <u>55</u> with potassium thiocyanate.
The key intermediate in this transformation is probably thiourea <u>55a</u>. Oxidative desulfurization occurs on treatment of
<u>56</u> with a mixture of sulfuric and nitric acids and the re-
sulting <u>57</u> is subjected to amide-ester interchange with anhydrous ethanolic hydrogen chloride to complete the synthesis of

<u>etomidate</u> (58).[21]

It is possible to form 2-imino-4-imidazolines, such as <u>59</u>, <u>in</u> <u>situ</u> from creatinine. Treatment of this heterocycle with 3-chlorophenylisocyanate leads to a sedative agent, <u>fenobam</u> (<u>60</u>).[22]

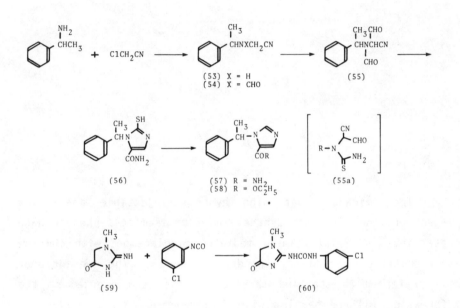

A substituted thiazole ring attached to a reduced imidazole moiety is present in a compound that displays antihypertensive activity. Reaction of thiourea <u>61</u> with methyl iodide to

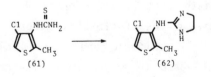

give the corresponding S-methyl analogue, followed by heating
with ethylenediamine, completes the synthesis of tiamenidine
(62).[23]

4. TRIAZOLES

Insertion of a triazole ring in place of an imidazole ring is
consistent in some cases with retention of antifungal activity.
The synthesis of one such agent, azoconazole (64), proceeds
simply by displacement of halide 63 with 1,2,4-triazole.[24] The
route to terconazole (65) is rather like that to ketoconazole
(34).[25]

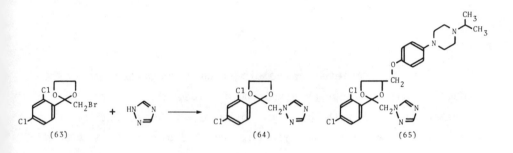

5. PYRAZOLINES

Reaction of substituted hydrazine analogue 66 with protected β-
dicarbonyl compound 67 leads to a ring-forming two-site reaction and formation of the pyrazoline diuretic agent, muzolimine
(68).[26]

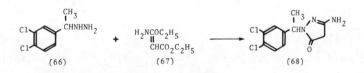

As a bioisoteric replacement for a substituted pyrrole
ring, a pyrazole ring is a key feature of the nonsteroidal

antiinflammatory agent, pirazolac (72). A Japp-Klingemann
reaction between 4-fluorobenzenediazonium chloride and ethyl
chloroacetate gives hydrazone 69. This is condensed with the
morpholino-enamine of p-chlorophenylacetaldehyde to give the
corresponding 4,5-dihydropyrazole 70. Treatment with hydrogen
chloride gives an eliminative aromatization reaction (71). The

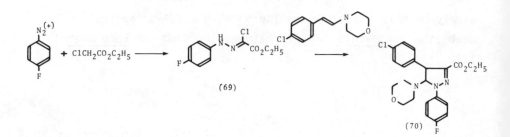

synthesis is completed by homologation through sequential reduction with diisopropylaluminum hydride, converstion to the
primary bromide with hydrogen bromide, displacement of that
function with potassium cyanide, and hydrolysis to the acid,
pirazolac (72), with potassium hudroxide in dimethyl sulfoxide.[27]

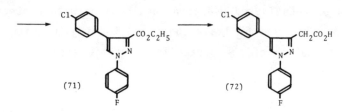

6. ISOXAZOLE

The 2-aminooxazole analogue, isamoxole (74), is an antiasthmatic agent. Its synthesis follows the classic pattern of condensation of hydroxy acetone with n-propylcyanamide to establish
the heterocyclic ring (73). The synthesis of isamoxole (74)

concludes by acylation with isopropyl chloride.[28]

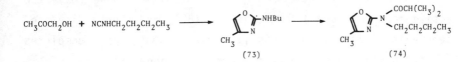

CH_3COCH_2OH + $NCNHCH_2CH_2CH_2CH_3$ ⟶

(73)   (74)

7. TETRAZOLES

Conversion of m-bromobenzonitrile to the tetrazole and addition of the elements of acrylic acid gives 75, starting material for the patented synthesis of the antiinflammatory agent, bropera-mole (76). The synthesis concludes by activation with thionyl chloride and a Schotten-Baumann condensation with piperidine.[29]

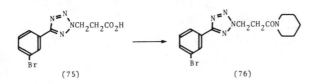

(75)   (76)

8. MISCELLANEOUS

Ropitoin (79) is an antiarrythmic compound containing a hydan-toin ring. Its synthesis is accomplished by alkylating 77 with chloride 78 with the aid of sodium methoxide.[30]

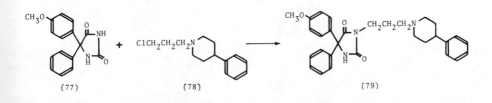

(77)   (78)   (79)

Reaction of ethyl cyanoacetate with ethyl thiolacetate produces a _Z_ and _E_ mixture of the dihydrothiazole derivative 80. This is _N_-alkylated with methyl iodide and base (81), the active methylene group is brominated (82), and then a displacement with piperidine (83) is performed. Hydrolysis completes the synthesis of the diuretic agent, ozolinone (84).[31]

Finally, a mesoionic sydnone, molsidomine (88), is active as an antianginal agent. Its synthesis starts by reacting 1-aminomorpholine with formaldehyde and hydrogen cyanide to give 85. Nitrosation gives the _N_-nitroso analogue (86) which

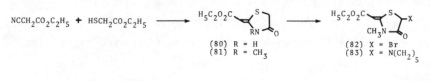

(80) R = H
(81) R = CH$_3$

(82) X = Br
(83) X = N(CH$_2$)$_5$

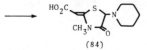

(84)

cyclizes to the sydnone (87) on treatment with anhydrous acid. Formation of the ethyl carbonate with ethyl chlorocarbonate completes the synthesis of molsidomine (88).[32]

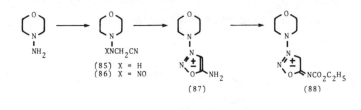

(85) X = H
(86) X = NO

(87)

(88)

REFERENCES

1. Y. L. L'Italien and T. C. Nordin, U.S. Patent 4,145,347 (1979); Chem. Abstr., 91, 39332p (1979).

2. M. A. Ondetti, B. Rubin, and D. W. Cushman, Science, 196, 441 (1977); Anon., Belgian Patent 851,361.

3. J. R. Carson and S. Wong, J. Med. Chem., 16, 172 (1973).

4. R. E. Johnson, Belgian Patent 16,542 (1974); Chem. Abstr., 83, 97007g (1975).

5. S. S. Pelosi, Jr., U.S. Patent 3,962,284 (1976); Chem. Abstr., 85, 159860g (1976).

6. H. R. Snyder, Jr., C. S. Davis, R. K. Bickerton, and R. P. Halliday, J. Med. Chem., 10, 807 (1967).

7. S. S. Petosi, Jr., R. E. White, R. L. White, G. C. Wright, and C.-N. You, U.S. Patent 3,919,231 (1975); Chem. Abstr., 84, 59168y (1976).

8. Anon., French Patent 2,384,765 (1980); Chem. Abstr., 92, 58595p (1980).

9. M. Hoffer and E. Grunberg, J. Med. Chem., 17, 1019 (1974).

10. A. G. Beaman and W. P. Tantz, U.S. Patent 3,865,823 (1975); Chem. Abstr., 82, 170944w (1975).

11. C. Jeanmart and M. N. Messer, German Offen. 2,035,573 (1969); Chem. Abstr., 74, 100044p (1971).

12. J. Heeres, L. J. J. Backx, J. H. Mostmans, and J. van Cutsem, J. Med. Chem., 22, 1003 (1979).

13. E. F. Godefroi, J. Heeres, J. Van Cutsem, and P. A. J. Janssen, J. Med. Chem., 12, 784 (1969).

14. G. E. Gymer, U.S. Patent 4,062,966.

15. K. A. M. Walker and M. Marx, U.S. Patent 4,038,409 (1976); Chem. Abstr., 87, 152210c (1977).

16. J. Heeres, U.S. Patent 3,936,470.

17.   J. Heeres, German Offen. 2,602,770 (1976); Chem. Abstr.,
      86, 29811b (1977).

18.   K. A. M. Walker, A. C. Braemer, S. Hitt, R. E. Jones, and
      T. R. Matthews, J. Med. Chem., 21, 840 (1978).

19.   T. H. Brown, G. J. Durant, J. C. Emmett, and C. R.
      Ganellin, U.S. Patent 4,145,546 (1979); Chem. Abstr., 90,
      204137t (1979).

20.   R. R. Crenshaw, G. Kavadias, and R. F. Santonge, U.S.
      Patent 4,157,340 (1979); Chem. Abstr., 91, 157312e (1979).

21.   E. F. Godefroi, P. A. J. Janssen, C. A. M. Van der Eycken,
      A. H. M. T. Van Heertum, and C. J. E. Niemegeers, J. Med.
      Chem., 8, 220 (1965).

22.   C. R. Rasmussen, U.S. Patent 3,983,135 (1976); Chem.
      Abstr., 86, 55441a (1977).

23.   H. Rippel, H. Ruschig, E. Lindner, and M. Schorr, German
      Offen. 1,941,761 (1971); Chem. Abstr., 74, 100054s (1971).

24.   G. Van Reet, J. Heeres, and L. Wals, German Offen.
      2,551,560 (1976); Chem. Abstr., 85, 94368f (1976).

25.   J. Heeres, L. J. J. Backx, and J. H. Mostmans, German
      Offen. 2,804,096 (1978); Chem. Abstr., 89, 180014b (1978).

26.   E. Moeller, K. Meng, E. Wehinger, and H. Horstmann, German
      Offen. DE 2,319,278 (1974); Chem. Abstr., 82, 57690x
      (1975).

27.   H. Biere, E. Schroder, H. Ahrens, J.-F. Kapp, and I.
      Bottcher, Eur. J. Med. Chem., 17, 27 (1982).

28.   W. J. Ross, R. G. Harrison, M. J. R. Jolley, M. C.
      Neville, A. Todd, J. P. Berge, W. Dawson, and W. J. F.
      Sweatman, J. Med. Chem., 22, 412 (1979).

29.   Anon., British Patent 1,319,357 (1973); Chem. Abstr., 79,
      92231h (1973).

30. S. Hayao, H. J. Havera, and W. G. Stryker, U.S. Patent 4,006,232 (1977); Chem. Abstr., 87, 5968c (1977).

32. K. Masuda, T. Kamiya, Y. Imashiro, and T. Kaneko, Chem. Pharm. Bull., 19, 72 (1971).

9 Six-Membered Heterocycles

The five-membered heterocycles discussed in the preceding chapter more often than not constituted the pharmacophoric moieties of the drugs in question. Drugs based on six-membered rings, on the other hand, constitute a somewhat more diverse group. In many cases such as the dihydropyridines and the antibacterial pyrimidines, the ring system again provides the pharmacophore; this section, however is replete with agents in which the heterocyclic ring simply serves as surrogate for an aromatic ring.

1.PYRIDINES

Derivatives of anthranilic acid have a venerable history as nonsteroid antiinflammatory agents. It is thus not surprising that the corresponding derivatives in which phenyl is replaced by pyridinyl show much the same activity.

145

Drugs that are too highly hydrophilic are often absorbed rather poorly from the gastrointestinal tract. It is sometimes possible to circumvent this difficulty by preparing esters of such compounds so as to change their water lipid partition characteristics in order to enhance absorption. Once absorbed, the esters are cleaved by the numerous esterase enzymes in the bloodstream, releasing free drug.

Preparation of the first of these antiinflammatory prodrugs starts with the displacement of halogen on bromophthalide 2 by the anion of the nicotinic acid derivative 1. Reaction of the intermediate 3 with aniline 4 leads to formation of talniflumate (5).[1]

In much the same vein, the basic ester 7 can be obtained by reaction of the same chloroacid with morpholine derivative 6. Reaction with aniline 4 affords morniflumate (8).

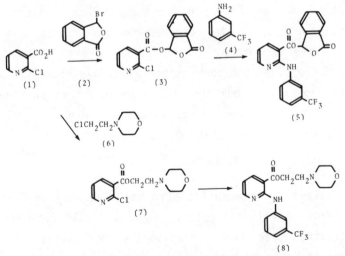

Congestive heart failure represents the end result of a

complex process which leads eventually to death when the
heart muscle is no longer able to perform its function as a
pump.   Cardiotonic agents such as <u>digitalis</u> (see Steroids)
have proved of value in treatment of this disease by
stimulating cardiac muscle.   The toxicity of these agents
has led to an extensive search for alternate drugs.   The
bipyridyl derivative <u>amrinone</u> has shown promise in the
clinic as a cardiotonic agent.

Starting material for the synthesis is the enaminoaldehyde <u>10</u>, obtainable by some version of the Villsmeyer
reaction on picoline derivative <u>9</u>.   Condensation of that
with cyanoacetamide in the presence of methoxide leads to
pyridone <u>12</u>.   The reaction can be rationalized by assuming
that the initial step consists in Michael addition of the
anion from acetamide to the acrolein; elimination of
dimethylamine would afford the intermediate <u>11</u>.   Condensation of amide nitrogen with the aldehyde leads to the
observed product.   Saponification of the nitrile then gives
acid <u>13</u>.   Treatment under nitrating conditions leads to an
interesting reaction that results ultimately in replacement
of the carboxyl by a nitro group (<u>14</u>).   Reduction of that
last function affords the amine; there is thus obtained
<u>amrinone</u> (<u>15</u>).[2]

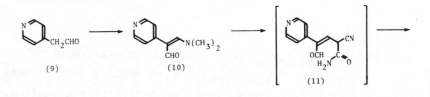

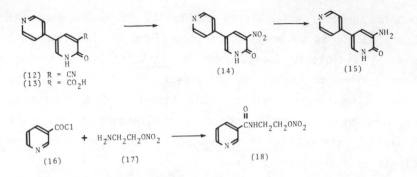

(12) R = CN
(13) R = CO₂H

(14)

(15)

(16)   +   H₂NCH₂CH₂ONO₂   ⟶
(17)

(18)

Organic nitrite esters such as nitroglycerin have constituted standard treatment for anginal attacks for the better part of the century. These agents, which are thought to owe their efficacy to reducing the oxygen demand of the heart, as a class suffer from poor absorption and very short duration of action. Nicordanil (18) represents an attempt to overcome these shortcomings by combining a nitrite with nicotinate, the latter moiety itself having some vaso-dilating action. The drug is obtained in straightforward manner by reaction of nicotinoyl chloride (16) with the nitrite ester (17) of ethanolamine.[3]

$$C_2H_5O_2CCH = C \begin{smallmatrix} CH_3 \\ \\ CH_3 \end{smallmatrix} + RO_2CCH_2CHCH_2C(CH_3)_2 \longrightarrow C_2H_5O_2CCH = C \begin{smallmatrix} CH_3 \\ \\ CH_2 \end{smallmatrix}$$

(19)   (20)   (21)

The excessive production of sebum associated with acne has made life miserable for many an adolescent. Research on acne has, as a rule, concentrated on therapy rather than prophylaxis. A pyrimidone forms an interesting exception, being described as an antiseborrheic agent. Starting keto-ester 21 can, at least in principle, be obtained by γacylation of the anion from acrylate 19 with a derivative of acid 20. Reaction with hydroxylamine under basic conditions would afford initially the oxime 22. This cyclizes to the N-hydroxypyridine, piroctone (23) under the reaction conditions.[4]

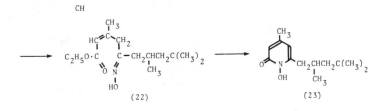

(22)   (23)

The so-called calcium channel blockers constitute a class of cardiovascular agents that have gained prominence in the past few years. These drugs, which obtund contraction of arterial vessels by preventing the movement of calcium ions needed for those contractions, have proved especially useful in the treatment of angina and hypertension. Dihydropyridines such as nifedipine (30) are particularly effective for these indications. A variation of the Hantsch pyridine synthesis used to prepare the parent molecule provides access to unsymmetrically substituted dihydropyridines. Though preparation of such compounds by ester interchange of but one ester has been described, these schemes are marked by lack of selectivity and low yields. Thus condensation of enaminoester 23 (obtained from the corresponding acetoacetate) with acetoacetate 24 and benzaldehyde 25 affords nimodipine (26).[5] In a similar

sequence, condensation of the enaminoester from methyl
acetoacetate (28) with acetoacetate 27 and benzaldehyde
gives the calcium channel blocker nicarpidine (29).[6]

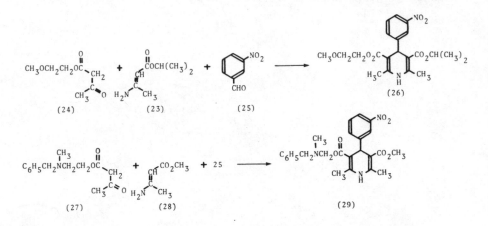

p-Fluorobutyrophenone derivatives of phenylpiperidines
constitute a class of very effective antipsychotic agents.
It is thus interesting to note that activity is retained
when a carbonyl group is inserted between phenyl and the
piperidine ring.   The starting benzoylpiperidine 32 can be
obtained by any of several schemes starting with the reduced
derivative of isonicotinic acid 31. Alkylation with bromo-
acetal 33 leads to the tertiary amine 34. Hydrolysis of the
acetal group leads to cloperone (35).[7]

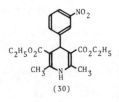

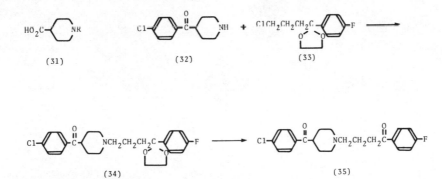

(31)   (32)   (33)

(34)   (35)

2. PYRIDAZINES

As noted earlier (Chapter 2) β-adrenergic blocking agents
have found extensive use in the treatment of hypertension.
Drawbacks of these drugs include relatively slow onset of
action and efficacy in only about half the patients
treated.   In an effort to overcome these shortcomings,
considerable research has been devoted to β-blockers which
incorporate moieties associated with direct-acting vasodilating agents.   (On the other side of the coin, the β-
blocking moiety should control some of the side effects
characteristic of the vasodilators, such as increase in
heart rate.)   Condensation of ketoester 36 with hydrazine
affords the corresponding pyridazinone 37.   Reaction of that
with phosphorus oxychloride leads to the chloro derivative
38.   The target compound could then be obtained by first
reacting the phenol with epichlorohydrin to give epoxide
39.   Opening of the oxirane with tertiary butylamine would
then complete construction of the β-blocking side chain
(40).   Displacement of chlorine by hydrazine then affords
prizidilol (41).[8]

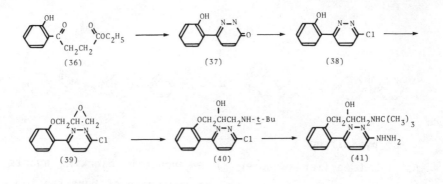

3.PYRIMIDINES

Though the great majority of antiinflammatory agents contain some form of acidic proton, occasional compounds devoid of such a function do show that activity. Thus the nonacidic pyrazolylpyrimidine epirazole (47) is described as a nonsteroid antiinflammatory agent. Reaction of pyrimidinone 42 with phosphorus oxychloride leads to the chloro derivative 43. Replacement of halogen with hydrazine gives the intermediate 44. Reaction of that with the methyl acetoacetate derivative 45 (obtained, for example, by pyro-lolysis of the orthoester) leads to formation of the pyra-zole ring. The reaction may be rationalized by assuming initial formation of hydrazone 46; addtion of the more basic hydrazine nitrogen to the masked carbonyl group followed by elimination of methoxide gives the observed product: there is thus obtained epirazole (47).[9]

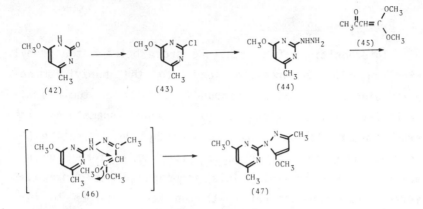

Two closely related diaminopyrimidines have been described as antineoplastic agents. In the absence of specific references, one may speculate that these can be prepared by a general method for the synthesis of aryl diaminopyrimidines.[10] Thus acylation of arylacetonitrile 48 with ethyl acetate affords the corresponding cyanoketone (49). Reaction of that intermediate with guanidine can be visualized as first involving formation of the imine derivative 50; addition of a second amino group from guanidine to the nitrile gives the cyclized derivative 51; tautomerization then gives the observed product, metoprine (52). The same sequence starting with ethyl propionate instead of ethyl acetate will lead to etoprine (53).

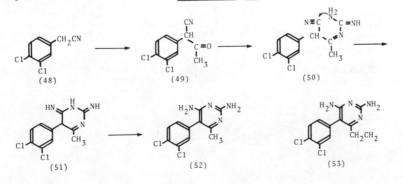

Interposition of a methylene group between the phenyl
ring and the heterocycle leads to the benzyldiamino-
pyrimidines, a class of compounds notable for their antibacterial activity. Condensation of hydrocinnamate 54 with
ethyl formate leads to the hydroxymethylene derivative 55.
In this case, too, the heterocyclic ring is formed by reaction with guanidine. This sequence probably involves
initial addition-elimination to the formyl carbon to form
56; cyclization in this case involves simple amide formation. Tautomerization then affords the hydroxy derivative
57.   This is converted to tetroxoprim (58) by first
replacing the hydroxyl by chlorine and then· displacement of
halogen with ammonia.[11]

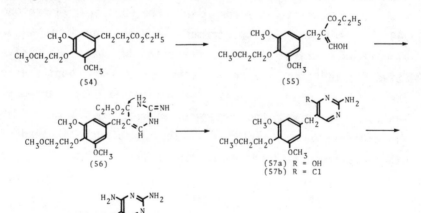

Preparation of the analogue <u>metioprim</u> involves an alternate approach. Aldol condensation of aldehyde <u>59</u> with propionitrile <u>60</u> gives the cinnamonitrile <u>61</u>. Reaction of this intermediate with guanidine probably involves displacement of the allylic aniline group as the first step (<u>62</u>). Cyclization followed by tautomerization affords <u>metioprim</u> (<u>63</u>).[12]

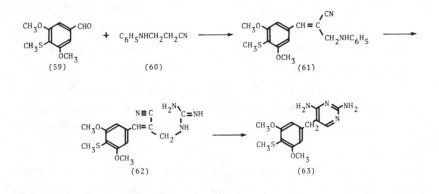

The pyrimidine <u>5-fluorouracil</u> (<u>64</u>) is used extensively in the clinic as an antimetabolite antitumor agent. As a consequence of poor absorption by the oral route, the drug is usually administered by the intravenous route. A rather simple latentiated derivative, <u>tegafur</u> (<u>66</u>), has overcome this limitation by providing good oral absorption. Reaction of 5-fluorouracil with trimethylsilyl chloride in the presence of base gives the disilylated derivative (<u>65</u>). Reaction of this with dihydrofuran (obtained by dehydrohalogenation of 2-chlorofuran) in the presence of stannic chloride affords directly tegafur (66).[13]

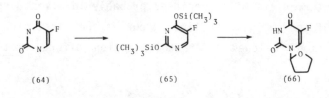

(64)   (65)   (66)

An aryloxypyrimidone has been described as an antiulcer agent; this activity is of note since the agent does not bear any structural relation to better known antiulcer drugs.   Displacement of halogen on the acetal of chloro-acetaldehyde by alkoxide from m-cresol gives the inter-mediate 67.   This affords enaminoaldehyde 68 when subjected to the conditions of the Villsmeyer reaction and subsequent hydrolysis.   Condensation with urea may be rationalized by assuming the first step to involve displacement of the dimethylamino group by an addition-elimination sequence (69).   Ring closure then leads to the pyrimidone and thus tolimidone (70).[14]

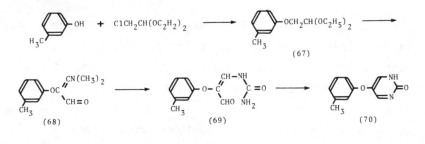

4.MISCELLANEOUS HETEROCYCLES

A cinnamoylpiperazine is described as an antianginal agent. The key intermediate 73 can, in principle, be obtained by alkylation of the monobenzyl derivative of piperazine 71 with ethyl bromoacetate (72). Removal of the protecting group then affords the substituted piperazine (73). Acylation of this with 3,4,5-trimethoxycinamoyl chloride gives cinepazet (74).[15]

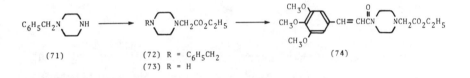

(71)   (72) R = $C_6H_5CH_2$ (74)
                (73) R = H

Though dental afflictions constitute a very significant disease entity, these have received relatively little attention from medicinal chemists. (The fluoride tooth-pastes may form an important exception.) This therapeutic target is, however, sufficiently important to be the focus of at least some research. A highly functionalized piperazine derivative that has come out of such work shows prophylactic activity against dental caries. Condensation of the enol ether 75 of thiourea with n-pentylisocyanate gives the addition product 77. Reaction of this with diamine 78, derived from piperazine, leads to substitution of the methylthio moiety by the primary amine, in all likelihood by an addition-elimination sequence. There is thus obtained ipexidine (79).[16]

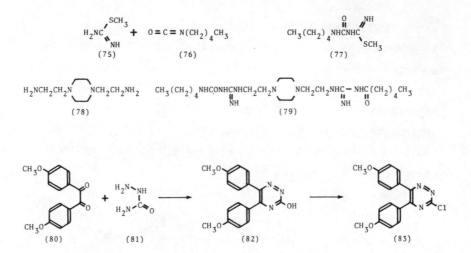

Heterocycles that carry p-anisyl groups on adjacent positions such as indoxole[17] and flumizole[17] constitute an important subclass among the nonsteroid antiinflammatory agents that do not possess an acidic proton. It is thus not very surprising to note that a similarly substituted 1,2,4-triazine also shows antiinflammatory activity. Condensation of the dibenzyl derivative 80 with semicarbazine affords the heterocyclic ring directly (82). Reaction with phosphorus oxychloride serves to convert the hydroxyl to chloro (83). Taking advantage of a reaction pioneered by Taylor, this intermediate is then reacted with an excess of the ylide from methyltriphenylphosphonium bromide. The first equivalent in all probability displaces halogen to form the substituted phosphonium salt 84. This is then converted to its ylide by excess phosphorane. Hydrolysis leads to loss of triphenylphosphine oxide. There is thus obtained anitrazafen (85).[18]

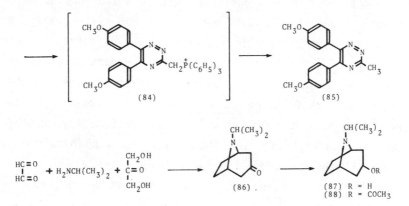

Acetylcholine is one of the fundamental neurotransmitters involved in a wide variety of normal regulatory functions. A number of disease states that may be associated with local excesses of this compound can at least in theory be treated by suppressing its action. Anticholinergic drugs have in practice proved of limited utility because it has been difficult to devise molecules that show much selectivity. The very widespread distribution of necessary cholinergic responses leads to the manifestation of a multitude of side effects when anticholinergic drugs are used in therapy. However, a number of syndromes could in principle, be treated with these drugs if they are applied by the topical route; lack of systemic absorption should avoid the side effects. It should, for example, be possible to treat the stomach lining to suppress the cholinergically mediated acid secretion associated with gastric ulcers. By the same token, direct administration to

the lung should prevent the bronchoconstriction associated
with asthma.   Considerable work based on this concept has
been occasioned by the observation that quaternary salts of
atropine (93), which should not be absorbed systemically, do
retain the anticholinergic activity of the parent base.   One
such salt, ipratropium bromide (92), has undergone considerable clinical investigation as an antiasthmatic agent administered by insufflation (i.e., topical application to the
bronchioles).   The fact that the stereochemistry of this
agent is the opposite from that which would be obtained by
direct alkylation with isopropyl bromide requires that a
somewhat longer sequence be employed for its synthesis.

     Preparation of the key tropine 86 is achieved by any
one of several variations on the method first developed by
Robinson, which involves reaction of a primary amine with
dihydroxyacetone and glyoxal.   Reduction of the carbonyl
group in the product (86) followed by acylation affords the
aminoester (88).   Transesterification with ester aldehyde 89

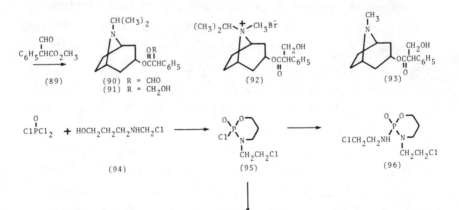

leads to **90**. The ester is then reduced to the atropic ester **91** by means of borohydride.[19] Attack of methyl bromide occurs from the more open face of the molecule to give ipratropium bromide (**92**).

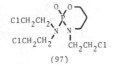

(100)   (97)   (98) R = H
                                                 (99) R = CH_2CH_2OH

Until the advent of the antitumor antibiotics, alkylating agents were the mainstay of cancer chemotherapy. The alkylating drug cyclophosphamide (**100**) found probably more widespread use than any other agent of this class. Two closely related agents, ifosfamide (**96**) and trofosfamide (**97**), show very similar activity; clinical development of these drugs hinges on the observation that the newer agents may show efficacy on some tumors that do not respond to the prototype. The common intermediate (**95**) to both drugs can be obtained from reaction of phosphorus oxychloride with amino alcohol **94**. Reaction of the oxazaphosphorane oxide with 2-chloroethylamine gives ifosfamide (**96**); displacement on bis(2-chloroethyl)amine gives trofosfamide (**97**).[20] In an alternate synthesis, the phosphorane is first condensed with the appropriate amino alcohols to give respectively **98** and **99**. These are then converted to the nitrogen mustards by reaction with mesyl chloride, or thionyl chloride.

REFERENCES

1.   M. A. Los, U.S. Patent 4,255,581; Chem. Abstr., 95, 62002x (1981).

2.   G. Y. Lesher and C. J. Opalka, Jr., U.S. Patent 4,107,315; Chem. Abstr. 90, 103844r (1979).

3.   H. Nagano, T. Mori, S. Takaku, I. Matsunaga, T. Kujirai, T. Ogasawara, S. Sugano, and M. Shindo, German Offen, 4,714,713; Chem. Abstr., 88, 22652h (1978).

4.   G. Lohaus and W. Dittmar, German Patent 1,795,831; Chem. Abstr., 89, 197347k (1978).

5.   E. Wehinger, H. Meyer, F. Bossert, W. Vater, R. Towart, K. Stoepel, and S. Kazada, German Offen, 2,935,451; Chem. Abstr. 95, 42922u (1981).

6.   Anon. Japanese Patent, 74,109,384; Chem. Abstr. 82, 170642c (1975).

7.   J. W. Ward and C. A. Leonard, French Demande 2,227,868; Chem. Abstr. 82, 170720v (1975).

8.   B. L. Lam, Eur. Pat. Appl. EP 47,164; Chem. Abstr., 96, 217866d (1982).

9.   Y. Morita, Y. Samejima, and S. Shimada, Japanese Patent 73 72,176; Chem. Abstr., 79, 137187s (1973).

10.   B. Roth and J. Z. Sterlitz, J. Org. Chem., 34, 821 (1969).

11.   W. Liebenow and J. Prikryl, French Demande 2,221,147; Chem. Abstr., 82, 156363z (1975).

12.   Anon. Belgian Patent 865,834; Chem. Abstr., 90, 54971u (1979).

13.   S. Hillers, R. A. Zhuk, A. Berzina, L. Serina, and A. Lazdins; U.S. Patent 3,912,734; Chem. Abstr., 84, 59538u (1976).

14. C. A. Lipinski, J. G. Stam, G. D. DeAngelis, and H. J. E. Hess; U.S. Patent 3,922,345; _Chem. Abstr._, _84_, 59552u (1976).

15. F. Faurau, G. Huguet, G. Raynaud, B. Pourias, and M. Turin; British Patent 1,168,108; _Chem. Abstr._, _72_, 12768 (1970).

16. R. A. Wohl; South African Patent 7,706,373; _Chem. Abstr._. _90_, 54980 (1979).

17. D. Lednicer and L. A. Mitscher, "The Organic Chemistry of Drug Synthesis", Vol. 2, Wiley, New York, 1980, p. 255

18. W. B. Lacefied and P. P. K. Ho, Belgian Patent 839,469; _Chem. Abstr._, _87_; 68431 (1977).

19. W. Schultz, R. Banholzer, and K. H. Pook, _Arzneim Forsch._ _26_, 960 (1976).

20. H. Arnold, N. Brock, F. Bourreaux, and H. Bekel, U.S. Patent 3,732,340; _Chem. Abstr._, _79_, 18772 (1973).

10 Five-Membered Heterocycles Fused to Benzene

1.INDOLES

Though the therapeutic utility of aspirin has been recognized for well over a century, this venerable drug was not classified as a nonsteroid antiinflammatory until recently. The first drug to be so classified was in fact, indomethacin (1), a very useful agent introduced into clinical practice within the past two-score years. Much of the work that led to the elucidation of the mechanism of action of this class of therapeutic agents was in fact carried out using indomethacin. This drug is often considered the prototype of cyclooxygenase (prostaglandin synthetase) inhibitors; it is still probably the most widely used inhibitor in various pharmacological researches. The undoubted good efficacy of the drug in the treatment of arthritis and inflammation at the same time has led to very widespread use in medical practice.

The relatively short duration of action of indo-methacin resulted in various attempts to develop prodrugs so as to overcome this drawback. One of these consists of an amino acid derivative. Thus, reaction of the drug with the chlorocarbonate derivative of dimethylethanol(2) affords the mixed anhydride 3. Reaction of that reactive intermediate with serine (4) leads directly to sermatacin (5).[1]

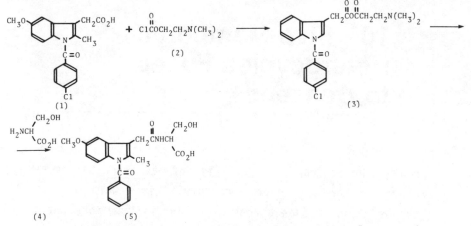

(1)   (2)   (3)

(4)   (5)

Replacement of chlorine on the pendant benzoyl group by azide is apparently consistent with antiinflammatory activity. Acylation of indomethacin intermediate 6 with p-nitrobenzoyl chloride leads to the corresponding amide (7). Saponification (8) followed by reduction of the nitro group gives the amine 9. The diazonium salt (10) obtained on treatment with nitrous acid is then reacted with sodium azide; there is thus obtained zidomethacin (11).[12]

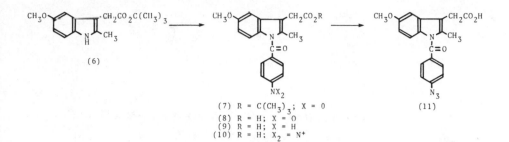

(6)

(7) R = C(CH$_3$)$_3$; X = O
(8) R = H; X = O
(9) R = H; X = H
(10) R = H; X$_2$ = N$^+$

(11)

Serotonin (12) is a ubiquitous endogenous compound that has a multitude of biological activities. For example, the compound lowers in certain biological tests. A compound that would lead to a serotonin derivative after decarboxylation has been described as an antihypertensive agent. (Note, however, that decarboxylation would have to occur by a mechanism different from the well-known biosynthetic loss of carbon dioxide from α-amino acids.) Mannich reaction on indole 13 with formaldehyde and dimethylamine gives the gramine derivative 14. Reaction with cyanide leads to replacement of the dimethylamino group to give the nitrile 15. Condensation of that intermediate with dimethyl carbonate and base gives the corresponding ester (16). Catalytic reduction of the nitrile group (17) followed by saponification affords indorenate (18).[3]

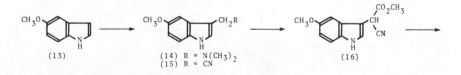

(13)

(14) R = N(CH$_3$)$_2$
(15) R = CN

(16)

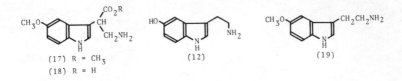

(17) R = CH$_3$
(18) R = H

(12)

(19)

Two closely related indoles fused to an additional saturated ring have been described as CNS agents. The first of these is obtained in straightforward manner by Fischer indole condensation of functionalized cyclohexanone **20** with phenylhydrazine (**19**). The product, cyclindole (**21**) shows antidepressant activity.[4] The fluorinated analogue flucindole (**26**) can be prepared by the same scheme. An alternate route starting from a somewhat more readily available intermediate involves as the first step Fischer condensation of substituted phenylhydrazine **22** with 4-hydroxycyclohexanone (**23**). The resulting alcohol (**24**) is then converted to its tosylate (**25**). Displacement by means of dimethylamine leads to the antipsychotic agent flucindole (**26**).[5]

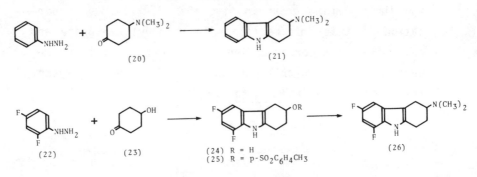

(20)　　　　　　　(21)

(22)　　　(23)　　　(24) R = H
　　　　　　　　　　(25) R = p-SO$_2$C$_6$H$_4$CH$_3$
　　　　　　　　　　　　　　　　(26)

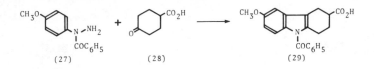

Changing the functionality on the alicyclic ring from an amine to a carboxylic acid leads to a compound that shows antiallergic activity, acting possibly by means of inhibition of the release of allergic mediators. Thus, condensation of acylated indole 27 with cyclohexanone carboxylic acid 28 affords directly oxarbazole (29).[6]

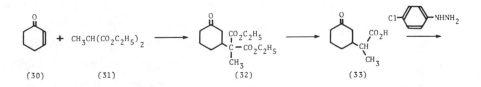

A fully unsaturated tricyclic indole derivative serves as the aromatic moiety for a nonsteroid antiinflammatory agent. Preparation of this compound starts with the Michael addition of the anion from methyl diethylmalonate to cyclohexanone. The product (32) is then hydrolyzed and decarboxylated to give ketoester 33. Fischer condensation with p-chlorophenylhydrazine leads to the indole 34. This is then esterified (35) and dehydrogenated to the carbazole 36. Saponification leads to the acid and thus carprofen (37).[7]

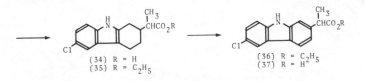

(34) R = H
(35) R = C_2H_5

(36) R = C_2H_5
(37) R = H

The salicylic acid functionality incorporated in a rather complex molecule interestingly leads to a compound that exhibits much the same activity as the parent. The 1,4 diketone required for formation of the pyrrole ring can be obtained by alkylation of the enamine from 2-tetralone (38) with phenacyl bromide. Condensation of the product, 39, with salicylic acid derivative 40 leads to the requisite heterocyclic system (41). The acid is then esterified (42) and the compound dehydrogenated to the fully aromatic system (43). Saponification affords fendosal (44).[8]

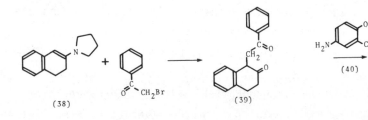

(38)   (39)   (40)

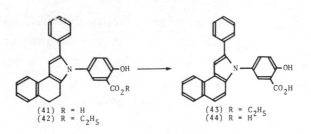

(41) R = H
(42) R = C$_2$H$_5$

(43) R = C$_2$H$_5$
(44) R = H

An isoindolinone moiety forms part of the aromatic moiety of yet another antiinflammatory propionic acid derivative.   Carboxylation of the anion from p-nitro-ethylbenzene (45) leads directly to the propionic acid (46).   Reduction of the nitro group followed by conden-sation of the resulting aniline (47) with phthalic anhydride affords the corresponding phthalimide (48). Treatment of that intermediate with zinc in acetic acid interestingly results in reduction of only one of the carbonyl groups to afford the isoindolone.   There is thus obtained indoprofen (49).[9]

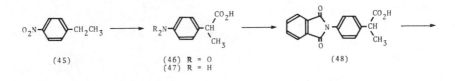

(45)

(46) R = O
(47) R = H

(48)

(49)

2.BENZIMIDAZOLES

A series of benzimidazole and benzimidazolone derivatives from the Janssen laboratories has provided an unusually large number of biologically active compounds, particularly in the area of the central nervous system. Reaction of imidazolone itself with isopropenyl acetate leads to the singly protected imidazolone derivative 51. Alkylation of this with 3-chloro-1-bromopropane affords the functionalized derivative 52. Use of this interintermediate to alkylate piperidine 53 (see cloperone, Chapter 6) affords the derivative 54. Hydrolytic removal of the isopropenyl group then gives the veterinary sedative milenperone (55).[10]The same sequence using p-fluorobenzoylpiperidine (56) gives the antipsychotic agent declenperone (57).[10]

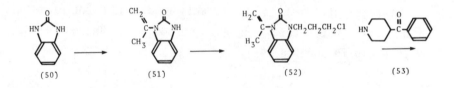

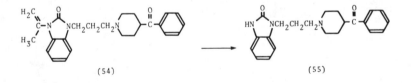

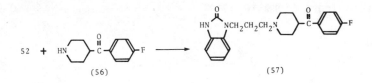

(56)   (57)

Alkylation of the monobenzhydryl derivative of piperazine (58) with the same alkylating agent gives oxatomide (59), after removal of the protecting group.[11] This agent shows antihistaminic activity as well as some mediator release inhibiting activity, a combination of properties particularly useful for the treatment of asthma.

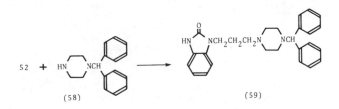

(58)   (59)

A somewhat more complex scheme is required for the preparation of benzimidazolones in which one of the nitrogen atoms is substituted by a 4-piperidyl group. The sequence starts with aromatic nucleophilic substitution on dichlorobenzene 60 by protected aminopiperidine derivative 61 to give 62. Reduction of the nitro group gives the diamine 63, which on treatment with urea affords the desired benzimidazolone 64.[12] The carbamate protecting group is then removed under basic conditions to give the secondary amine 65. Alkylation of this with the halide obtained by prior hydrolysis of

intermediate <u>52</u> affords <u>domperidone</u> (<u>66</u>), a very
promising antiemetic agent.[13]

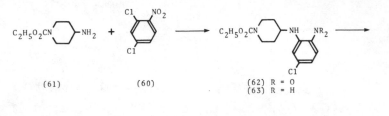

(61)   (60)   (62) R = O
                                                (63) R = H

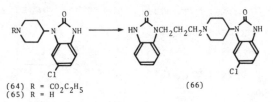

(64) R = CO$_2$C$_2$H$_5$ (66)
(65) R = H

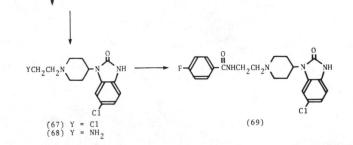

(67) Y = Cl   (69)
(68) Y = NH$_2$

Piperidinobenzimidazole <u>65</u> also serves as starting
material for the antipsychotic agent <u>halopemide</u> (<u>69</u>). In
the absence of a specific reference, one may speculate
that the first step involves alkylation with bromochloro-
ethane to give halide <u>67</u>. The chlorine may then be converted to the primary amine <u>68</u> by any of several methods
such as reaction with phthalimide anion followed by
hydrazinolysis. Acylation with <u>p</u>-fluorobenzoyl chloride
then gives the desired product.

A still different scheme is used for the preparation of the benzimidazole buterizine (74). Alkylation of benzhydrylpiperazine 58 with substituted benzyl chloride 70 gives the intermediate 71. Nucleophilic aromatic displacement on this compound by means of ethylamine leads to 72; reduction of the nitro group then gives the diamine 73. Treatment of that with the orthoformate ester of pentanoic acid serves to form the imidazole ring. There is thus obtained the peripheral vasodilating agent buterizine (74).[14]

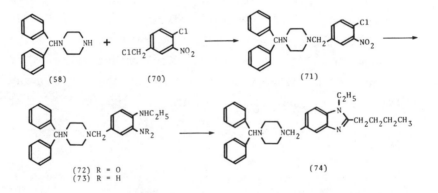

(58)   (70)   (71)

(72) R = O
(73) R = H   (74)

Amides and carbamates of 2-aminobenzimidazole have proved of considerable value as anthelminic agents, particularly in veterinary practice. A very considerable number of these agents have been taken to the clinic in the search for commercially exploitable agents. (See the section on Benzimidazoles in Chapter 11 of Volume 2 of this series.) A small number of additional compounds have been prepared in attempts to uncover superior agents.

In a typical synthesis, reduction of the nitro group in starting material 75 leads to the corresponding diamine 76. Reaction with intermediate 77 obtained by acylation of the methyl ether of thiourea with methyl chloroformate, leads directly to fenbendazole (78).[15]

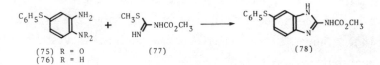

$$\begin{array}{cc}\text{(75)} & R = O \\ \text{(76)} & R = H\end{array} \qquad \text{(77)} \qquad \text{(78)}$$

Friedel-Crafts acylation of fluorobenzene with thiophene-1-carboxylic acid gives the ketone 79. Nitration proceeds ortho to the fluoro group to give the intermediate 80. Nucleophilic displacement by means of ammonia (81) followed by reduction of the nitro group leads to the corresponding amine 81. Treatment of that with reagent 77 gives the anthelmintic agent nocodazole (83).[16]

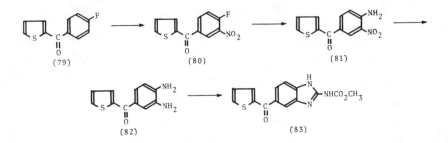

(79)   (80)   (81)

(82)   (83)

It is of particular note that slight changes in the functionality of this last-named compound lead to a profound change in biological activity.   The agent in question, <u>enviroxime,</u> shows pronounced antiviral activity. The synthesis of this compound begins with the reaction of diamine <u>84</u> with cyanogen bromide.   The reaction may be rationalized by assuming that cyanamide <u>85</u> is the initially formed product; addition of the remaining amine to the nitrile will give the observed product.   Reaction of the anion obtained on treatment of <u>86</u> with sodium hydride with isopropylsulfonyl chloride apparently affords <u>87</u> as the sole product.   (Note that minor tautomeric shifts could prodvide at least two alternate products.)   Reaction with hydroxylamine affords the <u>E</u> oxime as the predominant product.   There is thus obtained <u>enviroxime</u> (<u>88</u>).[17] Examination of the isomeric oximes show the <u>E</u> isomer to be a good deal more active than the <u>Z</u> counterpart.

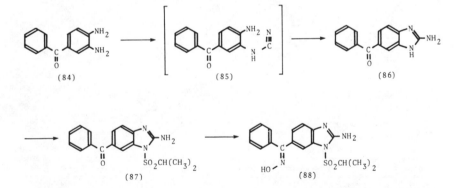

Incorporation of a 4-aminopiperidine moiety leads to a major change in biological activity.   The agent obtained by this modification, <u>astemizole</u> (<u>96</u>) is a rather potent antihistaminic compound.   Reaction of the isothiocyanate <u>89</u> with phenylenediamine under carefully

controlled conditions would lead to the thiourea 90.
Alkylation with p-fluorobenzyl bromide then leads to the
alkylated derivative 92.   Cyclization of that intermediate gives the benzimidazole 93.   The carbamate
protecting group is then removed under basic conditions.   Alkylation of the resulting secondary amine with
substituted phenethyl bromide 95 proceeds to give
astemizole (96).[18]

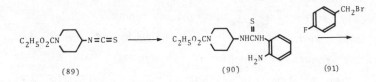

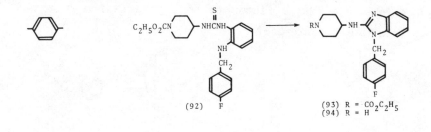

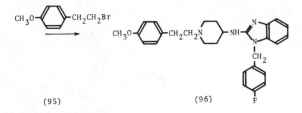

3.BENZOTHIAZOLES

Bioisosteric relations constitute one of the more
familiar tools in medicinal chemistry.   There are thus
sets of atoms that can often be interchanged without much

influence on the biological activity of the resulting molecules. In many series, for example, it may be quite useful to replace oxygen by sulfur; a sulfoxide sometimes serves in lieu of a ketone. Sulfur and nitrogen, on the other hand, are seldom considered to be a bioisosteric pair. It is thus of note that activity is retained in the antihelmintic compounds in the face of exactly such a substitution.

Reaction of aminothiophenol 97 with reagent 98 obtainable from phenylurea and thiophosgene leads directly to the anthelmintic agent frentizole (99).[19] In much the same vein, condensation of 100 with reagent 101 affords tioxidazole (102).

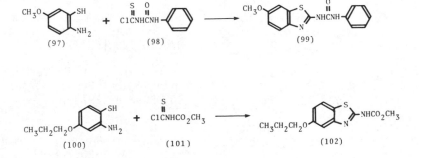

REFERENCES

1.   H. Biere, H. Arens, C. Rufer, E. Schroeder, and H. Koch, German Offen. 2,413,125; Chem. Abstr., 84, 17731 (1976)

2.   S. Tricerri, E. Panto, A. Bianchetti, G. Bossoni, and R. Venturini, Eur. J. Med. Chem., 14, 181 (1979)

3.   R. N. Schut, M. E. Safdy, and E. Hong, German Offen. 2,921,978; Chem. Abstr., 92, 128724f (1980).

4.   A. Mooradian, U.S. Patent 3,959,309; Chem. Abstr., 85, 123759s (1976).

5.   A. Mooradian, German Offen. 2,240,211 (1973); Chem. Abstr., 78, 136069x (1973).

6.   J. E. Alexander and A. Mooradian, U.S. Patent 3,905,998; Chem. Abstr., 84, 74101 (1976).

7.   L. Berger and A. J. Corraz, German Offen. 2,337,040; Chem. Abstr., 80, 108366g (1974).

8.   R. C. Allen and V. B. Anderson, U.S. Patent 3,931,457; Chem. Abstr., 84, 105391r (1976).

9.   G. Nannini, P. N. Giraldi, G. Malgara, G. Biasoli, F. Spinella, W. Logemann, E. Dradi, G. Zanni, A. Buttinoni, and A. Tommassini, Arzneim. Forsch. 23, 1090 (1973).

10. J. Vandenberk, L. E. J. Kennis, M. J. M. C. Van der Aa, and A. H. M. T. Van Heertum, German Offen. 2,645,125; Chem. Abstr., 87, 102326 (1977).

11. J. Vandenberk, L. E. J. Kennis, M. J. M. C. Van der Aa, and A. H. M. T. Van Heertum, German Offen. 2,714,437; Chem. Abstr., 88, 50920n (1978).

12. P. A. J. Janssen, A. H. M. T. Van Heertum, J. Vandenberk, and M. J. M. C. Van der Aa, German Offen. 2,257,261; Chem. Abstr., 84, 135657 (1976).

13. J. Vandenberk, L. E. J. Kennis, M. J. M. C. Van der Aa, and A. H. M. T. Van Heertum, German Offen. 2,632,870; Chem. Abstr., 87, 23274c (1977).

14. A. H. M. Raeymaekers, J. L. H. Van Gelder, G. M. Boeckx, and L. L. Van Hemeldonck, German Offen., 2,813,523; Chem. Abstr., 90, 54975y (1979).

15. H. Loewe, J. Urbanietz, R. Kirsch, and D. Duewel, German Offen. 2,164,690; Chem. Abstr., 79, 92217h (1973).

16.   J. L. H. Van Gelder, A. H. M. Raeymaekers, and L. F. C. Roevens, German Offen. 2,029,637; Chem. Abstr., 74, 100047s (1971).

17.   J. H. Wikel, C. J. Paget, D. C. DeLong, J. D. Nelson, C. Y. E. Wu, J. W. Paschal, A. Dinner, R. J. Templeton, M. O. Chaney, N. D. Jones, and J. W. Chamberlin, J. Med. Chem., 23, 368 (1980).

18.   F. Janssens, M. Luyckx, R. Stokbroekx and J. Torremans, U.S. Patent 4,219,559; Chem. Abstr., 94, 30579d (1981).

19.   C. J. Paget and J. L. Sands, German Offen. 2,003,841; Chem. Abstr., 73, 87920c (1970).

11 Benzofused Six-Membered Heterocycles

A diversity of biological effects are possessed by benzofused six-membered heterocycles. These range from antimicrobial activity to cardiovascular, CNS, and inflammation-influencing agents. It can be inferred that the ring system itself is primarily a molecular scaffold upon which to assemble the characteristic pharmacophore for the various receptors involved. It is interesting also to note that the range of bioactivities involved differ substantially from those seen with the benzofused five-membered heterocycles described in Chapter 10.

1. QUINOLINE DERIVATIVES

Various bioisosteric replacements for a phenolic hydroxyl have been explored. One such, a lactam NH, is incorporated into the design of the β-adrenergic blocker, carteolol (3). The fundamental synthon is carbostyril derivative 1. This is reacted in the usual manner with epichlorohydrin to give 2, which is in turn reacted with t-butylamine to complete the synthesis of carteolol (3), a drug that appears to have relatively reduced nonspecific myocardial depressant action.[1] Carrying this de-

183

vice farther results in the pseudocatechol, procaterol (6).

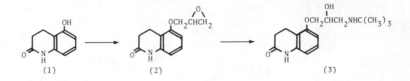

(1)   (2)   (3)

Friedel-Crafts alkylation of 8-hydroxycarbostyrils, such as 4,
leads to substitution at the C-5 position, namely, 5. In this
case an α-haloacyl reagent is employed. Displacement with isopropylamine and careful sodium borohydride reduction (care is

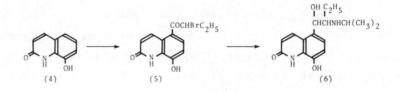

(4)   (5)   (6)

needed to avoid reduction of the carbostyril double bond) leads
to procaterol (6). Procaterol is an adren-ergic agonist selective for β$_2$-receptors. Thus it dilates bronchioles without
significant cardiac stimulation.[2]

N-aryl anthranilic acid derivatives ("fenamic acids")
often inhibit cyclooxygenase and thereby possess antiinflammatory and analgesic potency. One such agent, floctafenine
(9), can also be regarded as a 4-aminoquinoline. The synthesis
begins with a Gould-Jacobs reaction of m-trifluoromethylaniline
withdiethyl methoxymethylene malonate to give (after addition-
elimination and thermal cyclization) quinoline 7. Saponification and thermal decarboxylation gets rid of the now surplus
carbethoxy group. The phenolic OH is converted to a chloro
moiety with phosphorus oxychloride, which is displaced in turn

by methyl anthranilate to give fenamic acid __8__. This undergoes
ester exchange upon heating in glycerol to complete the
synthesis of prodrug __floctafenine__ (__9__).[3]

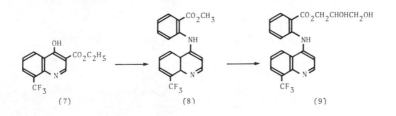

Interest in the antimicrobial properties of quinol-4-one-
3-carboxylic acids continues at a significant level. The synthesis of __rosoxacin__ (__12__) begins with a modified Hantsch pyridine synthesis employing as component parts ammonium acetate,
two equivalents of methyl propiolate, and one of 3-nitrobenz-
aldehyde. Oxidation of the resulting dihydropyridine (__10__) with
nitric acid followed by saponification, decarboxylation, and
reduction of the nitro group with iron and hydrochloric acid
gives aniline __11__. This undergoes the classic sequence of
Gould-Jacobs reaction with methoxymethylenemalonate ester to
form the 4-hydroxyquinoline ring, and then alkylation with
ethyl iodide and saponification of the ester to complete the
synthesis of the antibacterial agent __rosoxacin__ (__12__).[4]

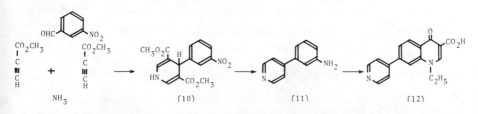

__Droxacin__ (__16__) is a carbabioisostere of the clinically
useful antimicrobial agent, __oxolinic acid__. Its synthesis

begins with nitrobenzofuran __13__ which is reduced with H_2 and a
palladized charcoal catalyst to give aniline __14.__ Gould-Jacobs
reaction with diethyl ethoxymethylenemalonate gives hydroxy-
quinoline __15__ along with some of the alternative cyclization
isomer. The synthesis is then completed in the usual way by
__N__-alkylation and saponification to __droxacin__ (__16__).[5]

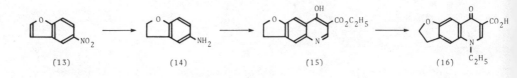

begins with nitrobenzofuran

An interestingly complex analogue in this family is __flume-__
__quine__ (__17__). As might be expected from the knowledge that the
bacterial target is an enzyme (DNAtopoisomerase II), one of the
enantiomers is quite potent but the other is not.[6]

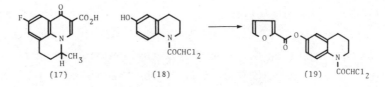

__Quinfamide__ (__19__) is one of a relatively small family of
antiamoebic compounds containing a dichloroacetamide function.
The synthesis begins by amidation of 6-hydroxytetrahydroquino-
line with dichloroacetyl chloride to give __18.__ The sequence is
completed by acylation with 2-furoyl chloride to give __quinf-__
__amide__ (__19__).[7]

2.   ISOQUINOLINE DERIVATIVES

__Nantradol__ (__25__) is an especially interesting agent in that it is
a potent analgesic that does not act at the morphine receptors.

It is quickly deacylated _in vivo_ and may qualify as a prodrug.
The published synthesis is rather long and bears conceptual
similarities to the synthesis of cannabinoids.   It has some
five asymmetric centers.   Dane salt formation between 3,5-di-
methoxyaniline and ethyl acetoacetate followed by borohydride
reduction gives synthon 20.   The amino group is protected by
reaction with ethyl chlorocarbonate, the ester group is saponified, and then cyclodehydration with polyphosphoric acid leads
to the dihydroquinolone ring system of 21.   Deblocking with HBr
is followed by etherification of the nonchelated phenolic hydroxyl to give 22.   Treatment with sodium hydride and ethyl
formate results in both N-formylation and C-formylation of the
active methylene to give 23.   Michael addition of methyl vinyl
ketone is followed by successive base treatments to remove the

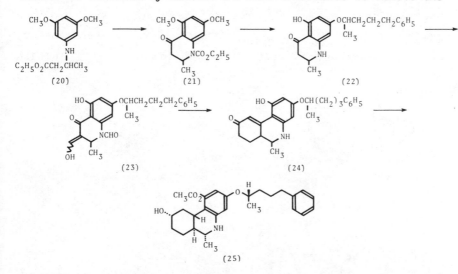

activating C-formyl group and then to complete the Robinson annulation to give 24.   Lithium in liquid ammonia reduces the
olefinic linkage and successive acetylation and sodium borohydride reductions complete the synthesis of nantradol (25).[8]

The l-form is much the more potent, being two to seven times
more potent than morphine as an analgesic. It is called levo-
nantradol.

3. BENZOPYRAN DERIVATIVES

The enzyme aldose reductase catalyzes the reduction of glucose
to sorbitol. Excess sorbitol is believed to contribute to cataracts and to neuropathy by deposition in the lens and nerves
of the eyes in the latter stages of diabetes mellitus. Spiro-
hydantoins have been found to inhibit this enzyme and so are of
potential value in preventing or delaying this problem. The S
enantiomers are the more potent. The synthesis of sorbinil
(32) illustrates a method developed for their chiral synthesis.
A chiral imine (28) is prepared by titanium tetrachloride-medi-
ated condensation of 6-fluorodihydrobenzopyran-4-one (26) with
S-α-methylbenzylamine (27) and this is reacted with hydrogen
cyanide to give 29 with a high degree of chirality transfer.
The basic nitrogen is next converted to the urea (30) with

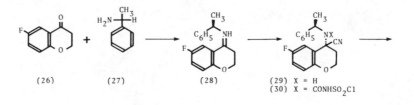

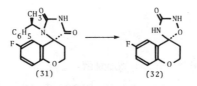

highly reactive chlorosulfonyl isocyanate. Treatment with
hydrogen chloride results in cyclization to the spirohydantoin

31 whose extraneous atoms are removed by hydrogen bromide treatment to give 4-(S)-sorbinil (32).[9]

Cannabinoids were used in medicine in the form of their crude extracts many centuries ago. Lately the use of cannabis for so-called recreational purposes has become a national vice of substantial proportions. Several attempts have been made to focus the potentially useful pharmacological properties of marijuana into drug molecules with no abuse potential.

Nabilone (37) is a synthetic 9-ketocannabinoid with anti-emetic properties. One of the best of the various published routes to nabilone starts with the enolacetate of nopinone (33), which on short heating with lead tetraacetate undergoes allylic substitution to give **34**. Treatment with p-toluene-sulfonic acid in chloroform at room temperature in the presence of the modified olivetol derivative **35** leads to condensation to **36**. Finally, treatment with stannic chloride at room temperature opens the cyclobutane ring and allows subsequent phenol capture to give optically active nabilone (37).[10]

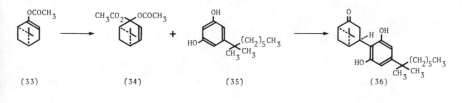

(33)   (34)   (35)   (36)

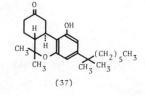

(37)

Nabitan (39) is a cannabis-inspired analgesic whose nitrogen atom was introduced in order to improve water solubility
and perhaps to affect the pharmacological profile as well. The
phenolic hydroxyl of benzopyran synthon 38 is esterified with
4-(1-piperidino)butyric acid under the influence of dicyclo-
hexylcarbodimide.[11] In addition to being hypotensive and
sedative-hypnotic, nabitran (39) is a more potent analgesic
than codeine. The preparation of synthon 38 begins with aceto-

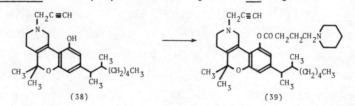

phenone 40, which undergoes a Grignard reaction and subsequent

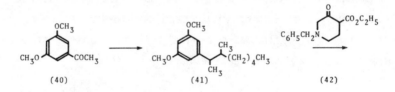

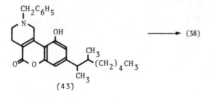

hydrogenolysis to put the requisite alkyl side chain in place
in 41. Ether cleavage (HBr/HOAc) is followed by condensation
with piperidone 42 to give tricyclic 43. Reaction with methyl-
magnesium bromide and hydrogenolysis of the benzylamine linkage
followed by alkylation gives 38.[12]

4. BENZODIOXANE DERIVATIVES

In the β-adrenergic blocking drug <u>pyrroxan</u> (<u>48</u>), the catechol moiety is masked in a doxane ring. The synthesis begins by alkylation of phenyl acetonitrile by 2-chloroethanol to produce alcohol <u>44</u>. Recuction converts this to amino alcohol <u>45</u> which undergoes thermal cyclization to 3-phenylpyrrolidine (<u>46</u>).

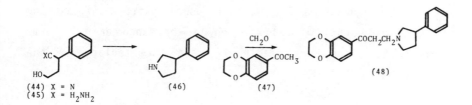

(44) X = N
(45) X = H₂NH₂   (46)   (47)   (48)

Finally, a Mannich reaction of <u>46</u> with formaldehyde and 4-acetyl-p-benzodioxane (<u>47</u>) leads to <u>pyrroxan</u> (<u>48</u>).[13]

5. BENZOXAZOLINONE DERIVATIVES

One of a variety of syntheses of the antipsychotic agent <u>brofoxine</u> (<u>50</u>) begins with a Grignard reaction on methyl anthranilate. The resulting product (<u>49</u>) is reacted with phosgene in pyridine and the synthesis is completed by bromination in acetic acid to give <u>brofoxine</u>.[14]

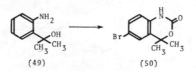

(49)   (50)

Another CNS active agent in this structural class is the tranquilizer-antidepressant <u>caroxazone</u> (<u>52</u>). Its published synthesis begins by reductive amination of salicylaldehyde and glycinamide to give <u>51</u>. The synthesis is completed by reaction with phosgene and sodium bicarbonate.[15]

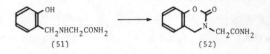

(51)   (52)

6. QUINAZOLINONE DERIVATIVES

The clinical acceptance of the dihydrochlorothiazide diuretics led to the synthesis of a quinazolinone bioisostere, fenquizone (54). The synthesis follows the usual pattern of heating anthranilamide (53) with benzaldehyde whereupon aminal formation takes place, presumably via the intermediacy of the Schiff's base.[16]

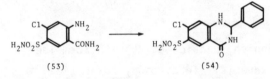

(53)   (54)

Synthesis of the CNS depressant/tranquilizer tioperidone (59) begins by alkylation of piperazine derivative 55 with 4-chlorobutyronitrile to give 56. Lithium aluminum hydride reduction gives primary amine 57, which is next reacted with isatoic anhydride to give anthranilamide analogue 58. Finally, reaction with phosgene gives tioperidone (59).[17]

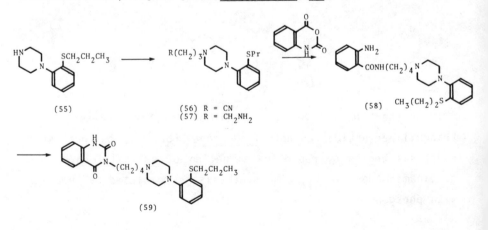

(55)   (56) R = CN
                        (57) R = CH$_2$NH$_2$ (58)

(59)

An apparently unexpected by-product of studies on 1,4-benzodiazepines is the antiinflammatory agent fluquazone (63). The synthesis begins by reaction of typical benzodiazepine synthon 60 with trifluoroacetyl chloride to give intermediate 61.   Reaction of this last with ammonium acetate leads to cyclization and cleavage to fluquazone (63).   This occurs, presumably, through a variant of a scheme involving facile cleavage of the labile trichloromethyl group, perhaps via 62, followed by cyclodehydration.[18]

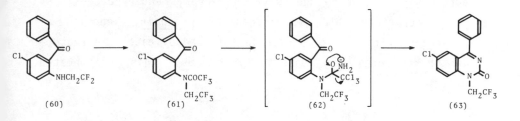

A related antiinflammatory agent prepared via a more traditional route is fluproquazone (65). Heating with urea in acetic acid results in transamidation by synthon 64 and subsequent cyclodehydration completes the synthesis.[19]

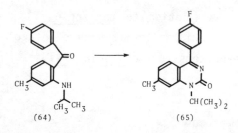

An antipsychotic agent with a chemical structure somewhat similar to that of tioperidone (59) is ketanserin (68). The synthesis involves the straightforward thermal alkylation of

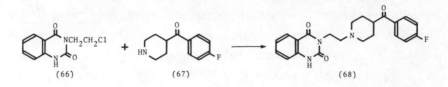

(66)   (67)   (68)

$\underline{N}_3$-(2-chloroethyl)quinazolinedione (66) with piperidinylketone 67.[20]

Alteration of the structural pattern produces a pair of adrenergic α-blocking agents which serve as antihypertensives. These structures are reminiscent of _prazocin_. Reaction of piperazine with 2-furoyl chloride followed by catalytic reduction of the furan ring leads to synthon 69. This, when heated

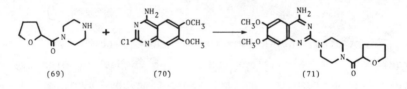

(69)   (70)   (71)

in the presence of 2-chloro-4-aminoquinazoline derivative 70, undergoes direct alkylation to _terazocin_ (71).[21] On the other hand, acylation of quinazoline 72 with oxadiazole derivative 73 gives the antihypertensive _tiodazocin_ (74).[22]

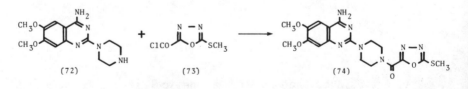

(72)   (73)   (74)

7. PHTHALAZINES

Phthalazines commonly possess adrenergic activity. One such, carbazeran (77), is a cardiotonic agent. Its patented synthesis involves nucleophilicaromatic displacement of chlorophthalazine derivative 75 with piperidinyl carbamate 76 to give carbazeran (77).[23]

8. BENZODIAZAPINES AND RELATED SUBSTANCES

The huge clinical success of drugs in this class has spawned an enormous list of congeners. Synthetic activity has, however, now slowed to the point that a separate chapter dealing with these heterocycles is no longer warranted.

Elfazepam (80) not only is a tranquilizer, but also stimulates feeding in satiated animals. One of several syntheses involves reaction of benzophenone derivative 78 with a glycine equivalent masked as an oxazolidine-2,5-dione (79).[24]

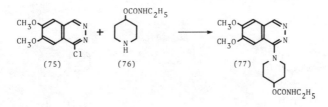

A water-soluble phosphine derivative of diazepam allows for more convenient parenteral tranquilizer therapy and avoids some complications due to blood pressure lowering caused by the propylene glycol medium otherwise required for administration. Fosazepam (82) is prepared from benzodiazepine 81 by sodium hydride-mediated alkylation with chloromethyldimethylphosphine oxide.[25]

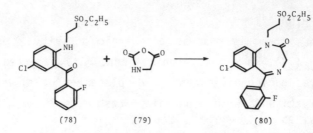

(78)   (79)   (80)

Lormetazepam (84) is readily synthesized by Polonovski rearrangement of benzodiazepine oxide derivative 83 by heating with acetic anhydride followed by saponification of the resulting rearranged ester.[26] The mechanism of this rearrangement to analogous tranquilizers has been discussed previously in this series.[27]

Quazepam (88) has a highly fluorinated sidechain so as to make this tranquilizer resistant to dealkylation. It also incorporates a lipid-solubilizing 2-thione moiety. The synthesis begins with biarylketone derivative 85 by N-alkylation with 2,2,2-trifluoroethyltriclate to give 86.

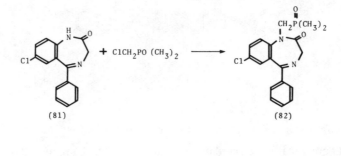

(81)   (82)

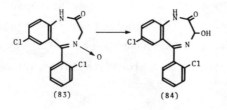

(83)   (84)

Next the product is acylated with bromoacetyl chloride and the glycine equivalent is constructed in place by a Gabriel amine synthesis (phthalamide anion followed by hydrazine) subsequent to which cyclization to benzodiazepine <u>87</u> occurs. The synthesis of the tranquilizer <u>quazepam</u> (<u>88</u>) is finished by thioamide conversion with phosphorus pentasulfide.[28]

A number of benzodiazepines have heterocyclic rings annelated to them. One such is the tranquilizer <u>midazolam</u> (<u>94</u>). Nitrosation (HONO) of secondary amine <u>89</u> leads to the <u>N</u>-nitroso analogue <u>90</u>. Nitrosoamidines, in the presence of carbanions, undergo carbon-carbon bond formation. Treatment of <u>90</u> with nitromethane and potassium <u>t</u>-butoxide results in formation of <u>91</u>. Raney nickel-catalyzed treatment reduces both the double bond and the alkyl nitro group to give saturated amine <u>92</u>. Treatment with either ethyl orthoacetate or acetic anhydride and polyphosphoric acid results in cyclization to <u>93</u> which is converted to the fused imidazole <u>94</u>, <u>midazolam</u>, on dehydrogenation with manganese dioxide.[29]

Another <u>alprazolam</u> (<u>95</u>) analogue is <u>adinazolam</u> (<u>98</u>). This

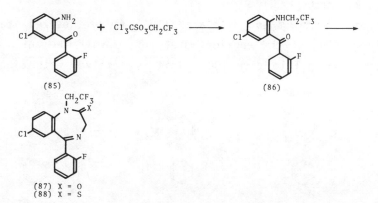

(85)

$+ Cl_3CSO_3CH_2CF_3 \longrightarrow$

(86)

(87) X = O
(88) X = S

substance is prepared from benzodiazepine synthon <u>96</u> by amidation of the hydrazine moiety with chloracetyl chloride followed by thermal cyclization in acetic acid to <u>97</u>. Reaction with potassium iodide and diethylamine results in net displacement of the allylic halogen and formation of the tranquilizer and antidepressant, <u>adinazolam</u> (<u>98</u>).[30]

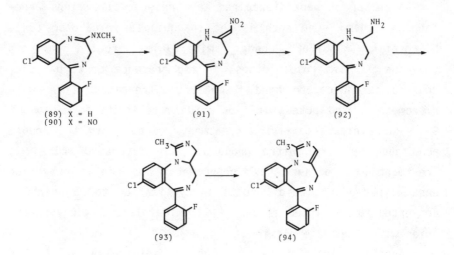

(89) X = H
(90) X = NO   (91)   (92)

(93)   (94)

9. MISCELLANEOUS

The antianginal agent <u>diltiazem</u> (<u>104</u>) is synthesized starting with opening of the epoxide moiety of <u>99</u> with the anion of 2-nitrothiophenol to give <u>100</u>. This is resolved with cinchoni-

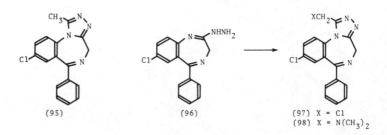

(95)   (96)   (97) X = Cl
                            (98) X = N(CH₃)₂

dine and reduced to the amine (101) before cyclodehydration to
lactam 102. This was alkylated with 2-chloroethyldimethyl-
amine, using dimethylsulfinyl sodium as base, to give 103. The
synthesis of the more active d-form of cardioactive diltiazem
(104) is concluded by acetylation with acetic anhydride and
pyridine.[31]

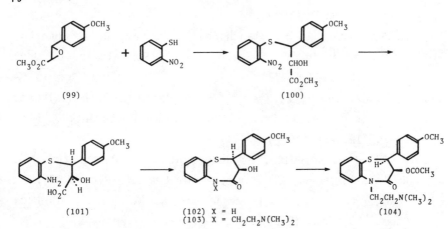

(99)   (100)

(101)   (102) X = H   (104)
               (103) X = $CH_2CH_2N(CH_3)_2$

REFERENCES

1. K. Nakagawa, N. Murakami, S.Yoshizaki, M. Tominaga, H.
    Mori, Y. Yabuuchi, and S. Shintani, J. Med. Chem., 17, 529
    (1974).

2. S. Yoshizaki, K. Tanimura, S. Tamada, Y. Yabuuchi, and K.
    Nakagawa, J. Med. Chem., 19, 1138 (1976).

3. A. Allais, Chim. Ther., 8, 154 (1973).

4. Y. Lescher and P. M. Carabateas, U.S. Patent 3,907,808
    (1975); Chem. Abstr., 84, 43880p (1975).

5. R. Albrecht, Eur. J. Med. Chem., 12, 231 (1977); Ann.,
    762, 55 (1972).

6. J. F. Gerster, S. R. Rohlfing, and R. M. Winandy, Abstr.
    N. Am. Med. Chem. Symp. 20-24 June, 1982, p. 153.

7.   D. M. Bailey, E. M. Mount, J. Siggins, J. A. Carlson, A.
     Yarinsky, and R. G. Slighter, J. Med. Chem., 22, 599
     (1979).

8.   M. R. Johnson and G. M. Milne, J. Heterocycl. Chem., 17,
     1817 (1980).

9.   R. Sarges, H. R. Howard, Jr., and P. R. Kelbaugh, J. Org.
     Chem., 47, 4081 (1982).

10. R. A. Archer, W. B. Blanchard, W. A. Day, D. W. Johnson,
     E. R. Lavagnino, C. W. Ryan, and J. E. Baldwin, J. Org.
     Chem., 42, 2277 (1977).

11. R. K. Razdan, B. Z. Terris, H. K. Pars, N. P. Plotnikoff,
     P. W. Dodge, A. T. Dren, J. Kyncl, and P. Somani, J. Med.
     Chem., 19, 454 (1976).

12. M. Winn, D. Arendsen, P. Dodge, A. Dren, D. Dunnigan, R.
     Hallas, H. Hwang, J. Kyncl, Y.-H. Lee, N. Plotnikoff, P.
     Young, H. Zaugg, H. Dalzell, and R. K. Razdan, J. Med.
     Chem., 19, 461 (1976).

13. V. A. Dobrina, D. V. Ioffe, S. G. Kuznetsov, and A. G.
     Chigarev, Khim. Pharm. Zh., 8, 14 (1974); Chem. Abstr.,
     81, 91445k (1974).

14. L. Bernardi, S. Coda, A. Bonsignori, L. Pegrassi, and G.
     K. Suchowsky, Experientia, 24, 774 (1968).

15. L. Bernardi, S. Coda, V. Nicolella, G. P. Vicario, A.
     Minghetti, A. Vigevani, and F. Arcamone, Arzneim. Forsch.,
     29, 1412 (1979); L. Bernardi, S. Coda, L. Pegrassi, and
     K. G. Suchowsky, Experientia, 24, 74 (1968).

16. G. Cantarelli, Il Farmaco, Sci. Ed., 25, 761 (1970).

17. R. F. Parcell, U.S. Patent 3,819,630; Chem. Abstr., 80;
     146190k (1974).

18. L. Bernardi, S. Coda, V. Nocolella, G. P. Vicario, A.
     Minghetti, A. Vigevani, and F. Arcamone, Arzneim. Forsch.,

29, 1412 (1979); L. Bernardi, S. Coda, L. Pegrassi, and
K. G. Suchowsky, Experientia, 24, 774 (1968).

19. P. G. Mattner, W. G. Salmond, and M. Denzer, French Patent
     2,174,828 (1973).

20. J. Vandenberk, L. E. J. Kennis, M. J. M. C. Van der Aa,
     and A. H. M. T. van Heertum, Eur. Patent Appl. 13,612
     (1980); Chem. Abstr., 94; 65718a (1981).

21. M. Winn, J. Kyncl, D. A. Dunnigan, and P. H. Jones, U.S.
     Patent 4,026,894 (1977); Chem. Abstr., 87; 68411m (1977).

22. R. A. Partyka and R. R. Crenshaw, U.S. Patent 4,001,237
     (1977); Chem. Abstr., 86; 140028r (1977).

23. S. F. Campbell, J. C. Danilewicz, A. G. Evans, and A. L.
     Ham, British Patent Appl. GB 2,006,136 (1979); Chem.
     Abstr., 91; 193331u (1979).

24. Anon., Japanese Patent JP 709,691 (1970); Chem. Abstr.,
     80; 133497r (1974).

25. E. Wolfe, H. Kohl, and G. Haertfelder, German Patent DE
     2,022,503 (1971); Chem. Abstr., 76; 72570c (1972).

26. S. C. Bell, R. J. McCaully, C. Gochman, S. J. Childress,
     and M. I. Gluckman, J. Med. Chem., 11, 457 (1968); S. C.
     Bell and J. C. Childress, J. Org. Chem., 27, 1691 (1962).

27. D. Lednicer and L. A. Mitscher, The Organic Chemistry of
     Drug Synthesis, Wiley, New York, Vol. 1, p. 365; Vol. 2,
     p. 402.

28. M. Steinman, J. G. Topliss, R. Alekel, Y.-S. Wong, and E.
     E. York, J. Med. Chem., 16, 1354 (1973).

29. A. Walser, L. E. Benjamin, Sr., T. Flynn, C. Mason, R.
     Schwartz, and R. I. Fryer, J. Org. Chem., 43, 936 (1978);
     R. I. Fryer, J. Blount, E. Reeder, E. J. Trybulski, and
     A. Walser, J. Org. Chem., 43, 4480 (1978).

30. J. B. Hester, Jr., A. D. Rudzik, and P. F. Von
    Voigtlander, J. Med. Chem., 23, 392 (1980).
31. H. Inoue, S. Takeo, M. Kawazu, and H. Kugita, Yakugaku
    Zasshi, 93, 729 (1973); H. Kugita, H. Inoue, M. Ikezaki,
    M. Konda, and S. Takeo, Chem. Pharm. Bull., 19, 595
    (1971).

12 Beta-Lactams

After 40 years of clinical use, benzylpenicillin (1) remains an extremely effective and useful drug for the treatment of infections caused by bacteria susceptible to it. It fails, however, to be a perfect drug on several grounds. Its activity spectrum is relatively narrow; it is acid and base unstable and so must be given by injection; increasingly strains carry enzymes (β-lactamases) that inactivate the drug by hydrolysis; and it is haptenic so that many patients become allergic to it. Many analogues have been synthesized in order to overcome these drawbacks and a substantial number of semisynthetic penicillins, cephalosporins, cephamycins, and so forth, have subsequently been marketed. Recently new impetus has been added to the field by the discovery of new ring systems in fermentation liquors and through development of novel synthetic approaches so that the field of β-lactam chemistry is now characterized by the feverish activity reflected in the number of entries in this chapter.

1. PENICILLINS

One of the most successful penicillin analogues has been

ampicillin (2). The relatively small chemical difference between ampicillin and benzylpenicillin not only allows for substantial oral activity but also results in a substantial broadening of antimicrobial spectrum so as to allow for use against many Gram-negative bacteria. Many devices have been employed in order to enhance still further the oral absorption of ampicillin. Bacampicillin (6) is a prodrug of ampicillin

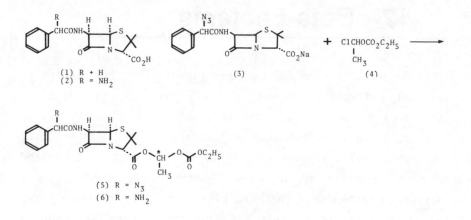

(1) R + H
(2) R = NH$_2$

(3)

(4)

(5) R = N$_3$
(6) R = NH$_2$

designed for this purpose. An azidopenicillin sodium salt (3) is reacted with mixed carbonate ester 4 (itself prepared from acetaldehyde and ethyl chlorocarbonate) to give ester 5. Reduction of the azido linkage with hydrogen and a suitable catalyst produces bacampicillin (6). Both enantiomers [starred (*) carbon] are active. The drug is rapidly and efficiently absorbed from the gastrointestinal tract and is quickly cleaved by serum esterases to bioactive ampicillin (2), acetaldehyde, carbon dioxide, and ethanol.[1]

Sarpicillin (10) is a double prodrug of ampicillin in that not only is the carboxy group masked as an ester, but a hetacillin-like acetonide has been added to the C-6 amide side chain. Its synthesis begins with the potassium salt of

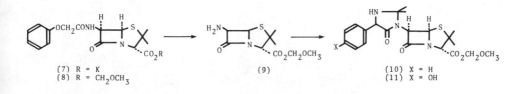

(7) R = K
(8) R = CH$_2$OCH$_3$

(9)

(10) X = H
(11) X = OH

penicillin <u>V</u> (<u>phenoxymethylpenicillin</u>, <u>7</u>) which is esterified with methoxymethyl chloride to give <u>8</u>. This is reacted with phosphorus pentachloride and the resulting imino chloride is cleaved to the free amine with <u>N,N</u>-dimethylaniline. Amine <u>9</u> then reacts with phenylglycyl chloride in acetone to complete the synthesis of <u>sarpacillin</u> (<u>10</u>).[2] The acetonide is presumably formed after acylation. A closely related prodrug of <u>amoxycillin</u>, known as <u>sarmoxicillin</u> (<u>11</u>), is made in the same way but with reaction of amine <u>9</u> with the hydrochloride of <u>p</u>-hydroxyphenylglycyl chloride in acetone being involved instead.[3]

An important molecular target of the β-lactam antibiotics is an enzyme that acts as a transpeptidase in the stepwise polymerization leading to a thickened, strong bacterial cell wall. Several amino acids are present in addition to the terminal <u>D</u>-alanyl-<u>D</u>-alanyl unit which the Strominger hypothesis suggests has the same overall shape and reactivity as <u>ampicillin</u>. This suggests that acylation of the amino group of <u>ampicillin</u> might lead to enhanced affinity or at least would be sterically allowable. It is interesting to find, therefore, that such acylation broadens the antimicrobial spectrum of the corresponding pencillins so that they now include the important

Gram-negative pathogen Pseudomonas aeruginosa.

Mezlocillin (13), one such agent, can be made in a variety of ways including reaction of ampicillin with chlorocarbamate 12 in the presence of triethylamine.[4]   Chlorocarbamate 12

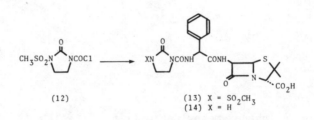

(12)

(13) X = SO$_2$CH$_3$
(14) X = H

itself is made from ethylenediamine by reaction with phosgene to form the cyclic urea followed by monoamide formation with methanesulfonyl chloride and then reaction of the other nitrogen atom with phosgene and trimethylsilyl chloride.   The closely related analogue azlocillin (14) is made in essentially the same manner as for mezlocillin, but with omission of the mesylation step.[5]   Interestingly, azlocillin is the more active of the two against many Pseudomonas aeruginosa strains in vitro.   An interesting alternative synthesis of azlocillin involves activation of the substituted phenylglycine analogue 15 with 1,3-dimethyl-2-chloro-1-imidazolinium chloride (16) and then condensation with 6-aminopenicillanic acid. [5,6]

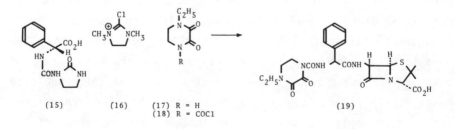

(15)   (16)   (17) R = H   (19)
                        (18) R = COCl

Another acylated ampicillin derivative with expanded antimicrobial spectrum is piperacillin (19). Its synthesis begins with 1-ethyl-2,3-diketopiperazine (17, which itself is made from N-ethylethylenediamine and diethyl oxalate), which is activated by sequential reaction with trimethylchlorosilane and then trichloromethyl chloroformate to give 18. This last reacts with ampicillin (2) to give piperacillin (19) which is active against, among others, the Enterobacteriaceae and Pseudomonads that normally are not sensitive to ampicillin.[7]

Continuing this theme, pirbenicillin (22) is another N-acylated antipseudomonal ampicillin analogue. Its synthesis begins by acylating 6-aminopenicillanic acid with N-carbobenzoxyphenylglycine by reaction with dicyclohexylcar-bodiimide and N-hydroxysuccinimide to activate the carboxyl group. The protecting CBZ group is removed from 19 on treatment with sodium carbonate to give 20. The synthesis of pirbenicillin is completed by reaction with 4-pyridoiminomethyl ether (21) (itself prepared from 4-cyanopyridine and anhydrous methanolic hydrogen chloride).[8]

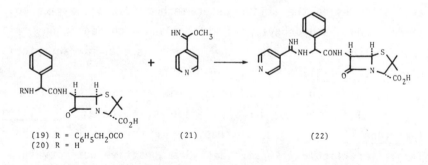

(19) R = C₆H₅CH₂OCO   (21)   (22)
(20) R = H

Piridicillin (27) is made by N-acylating amoxycillin with a rather complex acid. The synthesis begins by reacting N,N-diethanolamine with p-acetylbenzene-sulfonyl chloride to give 23. Conversion (to 24) with ethyl formate and sodium

methoxide is followed by base-catalyzed addition-elimination
with cyanoacetamide, during the course of which reaction
cyclodehydration occurs to produce the pyridone (25).
Saponification of the carboxylic acid 26 is followed by carboxy

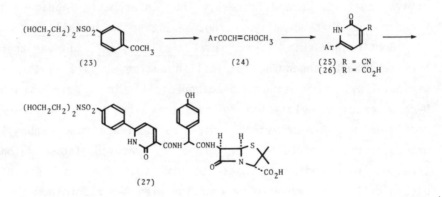

(23)   (24)   (25) R = CN
                                                (26) R = CO$_2$H

(27)

activation using the active ester method (dicyclohexylcarbodi-
imide and N-hydroxysuccinimide) and condensation with
amoxycillin to produce the broad spectrum antibiotic,
piridicillin (27).[9]

There is only one clinically significant penicillin at
present that does not have an amide side chain. Mecillinam
(amidinocillin, 29) has, instead, an amidine for a side chain.
It has very little effective anti-Gram positive activity but it
is quite effective against Gram-negative microorganisms. Its
synthesis begins by reacting N-formyl-1-azacycloheptane with
oxalyl chloride to form the corresponding imino chloride (28).
This is then reacted with 6-aminopenicillanic acid to produce
mecillinam.[10] A prodrug form, amidinocillin pivoxyl (30), is

made in the same manner by reaction of <u>28</u> with pivaloyloxy-
methyl 6-aminopenicillanoate instead.[11]

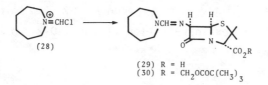

(28)

(29) R = H
(30) R = CH$_2$OCOC(CH$_3$)$_3$

2. CEPHALOSPORINS

Widespread clinical acceptance continues to be accorded to the
cephalosporins, and the field is extremely active as firms
search for the ultimate contender. Among the characteristics
desired is retention of the useful features of the older
members (relatively broad spectrum, less antigenicity than the
penicillins, relative insensitivity toward β-lactamases, and
convenience of administration) while adding better oral
activity and broader antimicrobial activity (particularly
potency against <u>Pseudomonas</u>, anaerobes, meningococci,
cephalosporinase-carrying organisms, and the like). To a
considerable extent these objectives have been met, but the
price to the patient has been dramatically increased.

<u>Cephachlor</u> (<u>35</u>) became accessible when methods for the
preparation of C-3 methylenecephalosporins became convenient.
The allylic C-3-acetoxyl residue characteristic of the natural
cephalosporins is activated toward displacement by a number of
oxygen- and sulfur-containing nucleophiles. Molecules such
as <u>31</u> can therefore be prepared readily. Subsequent reduction
with chromium(II) salts leads to the desired C-3
methylenecephems (<u>32</u>), which can in turn be ozonized at low
temperatures to produce the C-3 keto analogues. These are

isolated in the form of the C-3 hydroxycephem enolates (33).
Next, treatment with a variety of chlorinating agents (SOCl$_2$,
PCl$_3$, POCl$_3$, (COCl)$_2$, and COCl$_2$) in dry DMF solvent produces
the C-3 chloro analogues (34). The reaction can be carried out
so that the C-7 side chain is removed by the imino chloride
method so as to allow installation of the 7-D-2-amino-2-phenyl-
acetamido side chain of cephaclor (35).[12]

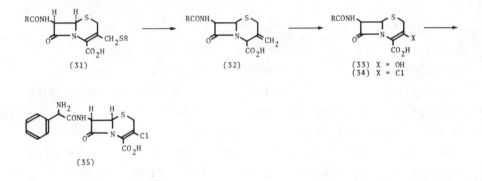

(31)   (32)   (33) X = OH
                                                (34) X = Cl

(35)

Conceptually closely related, cefroxadine (40) can be
prepared by several routes, including one in which the enol
(33) is methylated with diazomethane as a key step. A rather
more involved route starts with comparatively readily available
phenoxymethylpenicillin sulfoxide benzhydryl ester (36). This
undergoes fragmentation when treated with benzothiazole-2-thiol
to give 37. Ozonolysis (reductive work-up) cleaves the
olefinic linkage and the unsymmetrical disulfide moiety is
converted to a tosyl thioester (38). The enol moiety is
methylated with diazomethane, the six-membered ring is closed
by reaction with 1,5-diazabicyclo[5.4.0]undec-5-ene (DBU), and
the ester protection is removed with trifluoroacetic acid to

give <u>39</u>.   The amide side chain is removed by the usual
phosphorus pentachloride/dimethylaniline sequence followed by
reamidation with the appropriate acid chloride.   The result of
all this is <u>cefroxadine</u> (<u>40</u>).[13]

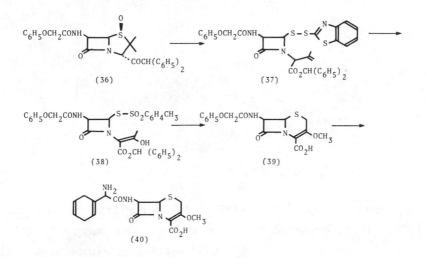

(36)   (37)   (38)   (39)   (40)

Possessing a side chain at C-7 reminiscent of that of
<u>amoxacillin</u> and a more typical sulfur containing C-3 moiety,
<u>cefatrizine</u> (<u>44</u>) can be synthesized by the active ester
mediated condensation of <u>t</u>-BOC-2-<u>p</u>-hydroxyphenylglycine (<u>41</u>)
with <u>N</u>-silyl 3-cephem synthon <u>42</u>.   The <u>t</u>-butyloxycarbonyl
protecting group of intermediate <u>43</u> is removed with formic acid
in order to complete the synthesis of <u>cefatrizine</u> (<u>44</u>), a broad
spectrum cephalosporin.[14] The thiol necessary for synthesis of
intermediate <u>42</u> from 7-aminocephalosporanic acid can be
prepared from 1-<u>N</u>-benzyl-2-azoimidazole (<u>45</u>) by lithiation
(<u>n</u>-butyllithium) followed by thiolation (hydrogen sulfide) to
give intermediate <u>46</u>.   The protecting benzyl moiety is then
removed reductively with sodium in liquid ammonia to give
synthon <u>47</u>.

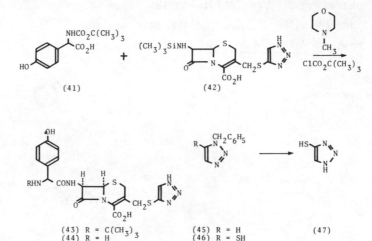

(41)   +   (42)

(43) R = C(CH$_3$)$_3$ (45) R = H   (47)
(44) R = H   (46) R = SH

Structurally rather similar to <u>cefatrizine</u> is <u>cefaparole</u> (<u>49</u>). It is prepared in quite an analogous manner by active ester condensation of <u>41</u> and 7-aminocephalosporanic acid analogue <u>48</u>. The blocking group is removed with trifluoro-acetic acid in anisole (9:1) to give <u>cefaparole</u>. Racemization does not take place during the synthetic sequence so the desired <u>R</u> stereochemistry of the side chain amino group is retained.[14]

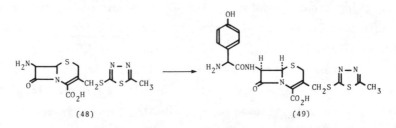

(48)   (49)

Analogous to <u>azlocillin-mezlocillin</u>, acylation of the amino group of 2-phenylglycine containing cephalosporins is consistent with antipseudomonal activity. There are many routes to cefoperazone (52). One of the more obvious is the condensation of cephalosporin antibiotic 50 with 2,3-diketo-piperazine 51 under modified Schotten-Baumann conditions.[15]

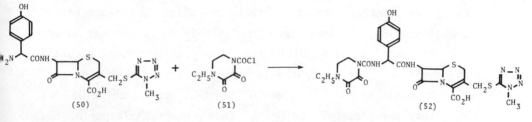

(50)   (51)   (52)

Cefonicid (55) is synthesized conveniently by nucleophilic displacement of the C-3 acetoxy moiety of 53 with the appropriately substituted tetrazole thiol (54).[16] The mandelic acid amide C-7 side chain is reminiscent of cefamandole.

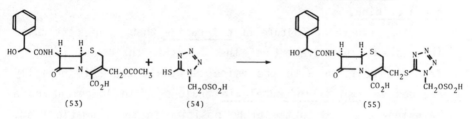

(53)   (54)   (55)

Cefazaflur (58) stands out among this group of analogues because it lacks an arylamide C-7 side chain (see cephacetrile for another example).[17] Cefazaflur (58) is synthesized by reaction of 3-(1-methyl-1H-tetrazol-5-ylthiomethylene)-7-amino-cephem-4-carboxylic acid (56) with trifluoromethylthioacetyl chloride (57).[18]

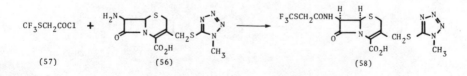

(57)   (56)   (58)

Cefsulodin (60) has a sulfonic acid moiety on the C-7 acyl side chain. This moiety conveys antipseudomonal activity to certain penicillins, and it is interesting to note that this artifice works with cefsulodin as well. It is also interesting that, in contrast to the other so-called third-generation cephalosporins, the spectrum of cefsulodin is rather narrow and its clinical success will place a premium upon accurate diagnosis. Its synthesis begins by acylation of 7-aminocephalosporanic acid with the acid chloride of 2-sulfonylphenylacetyl chloride to give cephalosporin 59. Reaction of that intermediate with aqueous potassium iodide and isonicotinic acid amide results in acetoxyl displacement from C-3 and formation of cefsulodin (60).[21] The quaternary base at C-3 is reminiscent of the substitution pattern of cephaloridine.

The structural feature of ceforanide that is of particular interest is the movement of the C-7 side chain amino moiety from the position α to the amide carbonyl, where it normally resides in ampicillin/cephalexin analogues, to lodgement on a methylene attached to the ortho position on the aromatic ring.

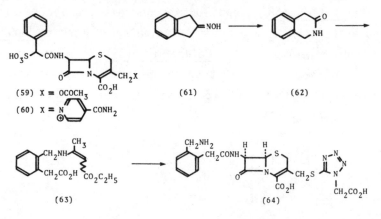

(59) X = OCOCH$_3$

(60) X = N⊕—pyridyl—CONH$_2$

(61)

(62)

(63)

(64)

This puts it geometrically in the same general area but considerably alters the electronic character of the molecule upon protonation. The synthesis of ceforanide (64) begins with a Beckmann rearrangement of the oxime of 2-indanone (61) to give lactam 62. Hydrolysis followed by protection of the amino group as the enamine (63) allows for subsequent mixed anhydride (isobutylchlorocarbonate)-mediated amide formation with the corresponding 7-aminocephalosporin synthon to give ceforanide (64). The requisite nucleophile for the C-3 moiety is prepared simply by carbonation of the lithio derivative of 1-methyl-1-H-tetrazol-5-ylthiol.[22]

Cefotiam (67) has an acyl aromatic C-7 side chain bioisosteric with an anilino ring. It can be prepared by acylation of the suitable acid moiety with 4-chloroacetoacetyl

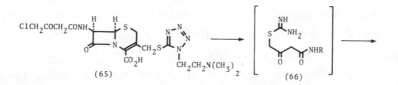

chloride to give amide 65. Chloride displacement with thiourea leads to cyclodehydration to the aminothiazole cefotiam (67), probably via intermediate 66.[23]

The interposition of a syn-oximino ether moiety between the amide carbonyl and the aromatic ring has proved richly

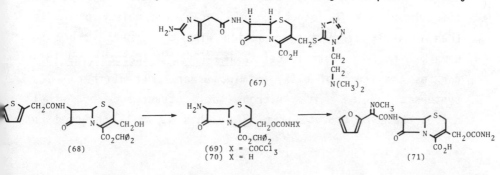

rewarding in that substantial resistance to β-lactamases results from this steric hindrance. A large number of analogues
now bear this feature, for example, cefuroxime (71). The
synthesis can be accomplished in a variety of ways. Benzhydryl
ester 68 (preparable from cephalothin[24]) is acylated with tri-
chloroacetyl isocyanate and the side chain at C-7 is removed
via the imino chloride method (pyridine and phosphorus pentachloride followed by tosic acid) to produce 7-aminocephalospor-
anic acid analogue 69. The carbamate moiety is partially
hydrolyzed to give 70 through use of anhydrous methanolic
hydrogen chloride (generated with methanol and acetyl
chloride). Next, the benzhydryl ester is cleaved with trifluoroacetic acid and the synthesis is concluded by a Schotten-
Baumann acylation with the appropriate syn-oximinoether-bearing
acid chloride so as to produce cefuroxime.[25]

     An analogous third-generation cephalosporin whose
synthesis illustrates one of the methods of preparing the
requisite oximino acid side chains is cefotaxime (76).
syn-Oxime 72 is methylated stepwise with dimethyl sulfoxide and
base and then chlorinated (Cl_2 in chloroform) to produce 73.
Alkylation of thiourea with this product results in concomitant
cyclodehydration to produce aminothiazole 74. The primary
amino group is then blocked with chloroacetyl chloride, the
ester group is saponified, and then the intermediate is used to
acylate 7-aminocephalosporanic acid in the usual way via the
acid chloride. The blocking chloroacetamide moiety of 75 is
then cleverly removed by reaction with thiourea in order to
unmask β-lactamase stable cefotaxime (76).[26] Ceftazidime (81)
carries the theme of bulky oximino ethers much further. Its
synthesis begins with nitrous acid treatment of ethyl
acetoacetate to produce oxime 77. This is next converted to

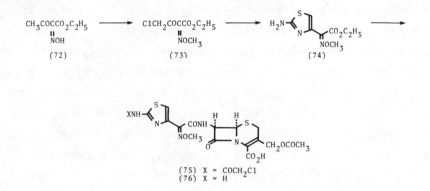

(72)   (73)   (74)

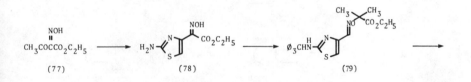

(75) X = COCH₂Cl
(76) X = H

a 2-aminothiazole (<u>78</u>) by halogenation with sulfuryl chloride
followed by thiourea displacement. The amino group is

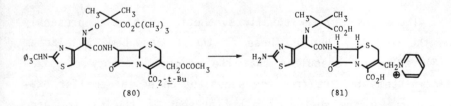

(77)   (78)   (79)

(80)   (81)

protected as the trityl amine and then ether formation with
ethyl 2-bromo-2-methylpropionate gives intermediate <u>79</u>.
Saponification next frees the carboxy group for condensation
with <u>t</u>-butyl 7-aminocephalosporinate mediated by dicyclohexyl-
carbodiimide and 1-hydroxybenzotriazole. The synthesis is
completed by removal of the protecting groups from <u>80</u> with trifluoroacetic acid and displacement of the acetoxyl moiety from
C-3 by treatment with pyridine and sodium iodide in order to
give <u>ceftazidime</u>. <u>Ceftazidime</u> (<u>81</u>) is quite resistant to β-
lactamases and possesses useful potency against pseudomonads.[25]

   <u>Ceftizoxime</u> (<u>83</u>) is structurally of interest in that it
lacks any functionality at C-3 and therefore cannot undergo the
usual metabolic deacetylation experienced by many cephalosporins. Its synthesis involves condensation of the acid chloride
corresponding to ester <u>74</u> (<u>74a</u>) with β-lactam <u>82</u>.[26]

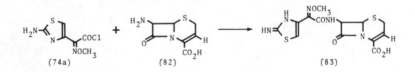

   There are very few totally synthetic antibiotics presently
on the market. One of these is the 1-oxacephem, <u>moxalactam</u>
(<u>96</u>). One may speculate that the enhanced potency of <u>moxa-
lactam</u> stems in part from the substitution of the smaller oxygen atom for the sulfur normally present in the six-membered
ring of cephalosporins thereby enhancing the reactivity of the
adjoining four-membered ring. It is also partly a measure of
the present stage of development of chemical synthesis and of
the relative economics of production of 7-aminocephalosporanic
acid that such an involved synthesis apparently is economically
competitive.

Pieces of various routes to moxalactam have been published
from which the following may be assembled as one of the plaus-
ible pathways.   The benzhydrol ester of 6-aminopenicillanic
acid (84) is S-chlorinated and treated with base whereupon the
intermediate sulfenyl chloride fragments (to 85).   Next, displacement with propargyl alcohol in the presence of zinc chloride gives predominantly the stereochemistry represented by diastereoisomer 86.   The side chain is protected as the phenyl-
acetylamide; the triple bond is partially reduced with a 5%
Pd-CaCO$_3$ catalyst and then epoxidized with m-chloroperbenzoic
acid to give 87.   The epoxide is opened at the least hindered
end with the lithium salt of 1-methyl-1H-tetrazol-5-ylthiol to
put in place the future C-3 side chain and give intermediate
88.   Jones oxidation followed in turn by ozonolysis (reductive
work-up with zinc-acetic acid) and reaction with thionyl
chloride and pyridine give halide 89.   The stage is now set for
an intramolecular Wittig reaction.   Displacement with triphenylphosphine and Wittig olefination gives 1-oxacephem 90.
Next a sequence is undertaken of side chain exchange and introduction of a C-7 methoxyl group analogous to that which is
present in the cephamycins and gives them enhanced β-lactamase
stability.   First 90 is converted to the imino chloride with
PCl$_5$ and then to the imino methyl ether (with methanol) and
next hydrolyzed to the free amine.   Imine formation with
3,5-di-t-butyl-4-hydroxybenzaldehyde is next carried out
leading to 91.   Oxidation with nickel peroxide gives
iminoquinone methide 92, to which methanol is added in a conjugate sense and in the stereochemistry illustrated in formula
93.   The imine is exchanged away with Girard reagent T to give
94, and this is acylated by a suitable protected arylmalonate,
as the hemiester hemiacid chloride, so as to give 95.   Deblocking with aluminum chloride and anisole gives moxalactam (96).

Moxalactam is a synthetic antibiotic with good activity against Gram-negative bacteria including pseudomonads and has excellent stability against β-lactamases.[27]

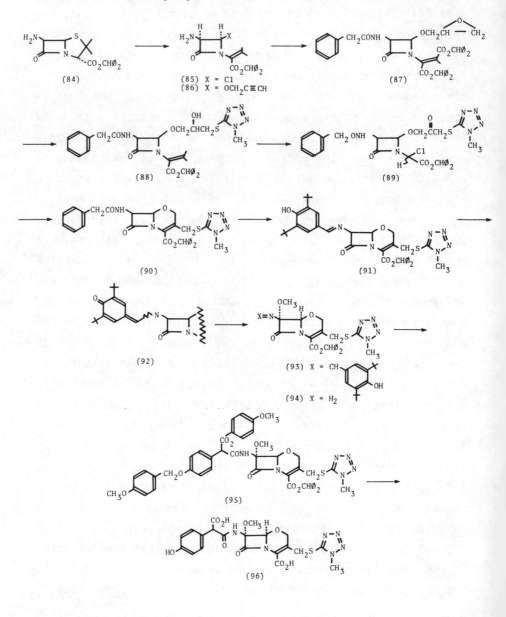

REFERENCES

1. B. A. Ekstrom, O. K. J. Kovacs, and B. O. H. Sjoberg,
    German Offen. 2,311,328 (1973); Chem. Abstr., 80, 14921q
    (1974).

2. P. Sleezer and D. A. Johnson, German Offen. 2,244,915
    (1973); Chem. Abstr., 78, 159590z (1973).

3. P. Sleezer and D. A. Johnson, South African 75 04,722
    (1976); Chem. Abstr., 86, 171440y (1977).

4. W. Schroeck, H. R. Furtwaengler, H. B. Koenig, and K. G.
    Metzer, German Offen. 2,318,955 (1973); Chem. Abstr., 82,
    31313b (1975).

5. H. B. Koenig, K. G. Metzer, H. A. Offe, and W. Schroeck,
    Eur. J. Med. Chem., 17, 59 (1982).

6. V. J. Bauer and S. R. Safir, J. Med. Chem., 9, 980 (1966).

7. I. Saikawa, S. Takano, C. Yoshida, O. Takashima, K.
    Momonoi, T. Yasuda, K. Kasuya, and M. Komatsu, Yakugaku
    Zasshi, 97, 980 (1977).

8. E. S. Hamanaka and J. G. Stam, South African 74 00,509
    (1973); Chem. Abstr., 83, 58808z (1975)

9. J. S. Kaltenbronn, J. H. Haskell, L. Daub, J. Knoble, D.
    DeJohn, U. Krolls, N. Jenesel, G.-G. Huang, C. L. Heifitz,
    and M. W. Fischer, J. Antibiotics, 32, 621 (1978).

10. F. Lund and L. Tybring, Nature, New Biol., 236, 135
    (1972).

11. F. J. Lund, German Offen. 2,055,531 (1971); Chem. Abstr.,
    75, 49070k (1971).

12. R. R. Chauvette and P. A. Pennington, J. Med. Chem., 18,
    403 (1975).

13. R. B. Woodward and H. Bickel, U.S. Patent 4,147,864
    (1979); Chem. Abstr., 91, 74633j (1979).

14. G. L. Dunn, J. R. E. Hoover, D. A. Berges, J. J. Taggart,

L. D. Davis, E. M. Dietz, D. R. Jakas, N. Yim, P. Actor, J. V. Uri and, J. A. Weisbach, J. Antibiotics, 29, 65 (1976).

15. I. Saikawa, S. Takano, Y. Shuntaro, C. Yoshida, O. Takashima, K. Momonoi, S. Kuroda, M. Komatsu, T. Yasuda, and Y. Kodama, German Offen., DE 2,600,880 (1977); Chem. Abstr., 87, 184533b (1977).

16. D. A. Berges, U.S. Patent 4,093,723 (1978); Chem. Abstr., 89, 180025f (1978).

17. D. Lednicer and L. A. Mitscher, The Organic Chemistry of Drug Synthesis, Wiley, New York, 1980, Vol. 2, p. 441.

18. R. M. deMarinis, J. C. Boehm, G. L. Dunn, J. R. E. Hoover, J. V. Uri, J. R. Guarini, L. Philips, P. Actor, and J. A. Weisbach, J. Med. Chem., 20, 30 (1977).

19. H. Nomura, T. Fugono, T. Hitaka, I. Minami, T. Azuma, S. Morimoto, and T. Masuda, Heterocycles, 2, 67 (1974).

20. W. J. Gottstein, M. A. Kaplan, J. A. Cooper, V. H. Silver, S. J. Nachfolger, and A. P. Granatek, J. Antibiotics, 29, 1226 (1979).

21. M. Namata, I. Minamida, M. Yamaoka, M. Shiraishi, T. Miyawaki, H. Akimoto, K. Naito, and M. Kida, J. Antibiotics, 31, 1262 (1978).

22. D. Lednicer and L. A. Mitscher, The Organic Chemistry of Drug Synthesis, Wiley, New York, 1977, Vol. 1, p. 417, 420.

23. M. C. Cook, G. I. Gregory, and J. Bradshaw, German Offen. DE2,439,880 (1975); Chem. Abstr., 83, 43354z (1975).

24. M. Ochiai, A. Morimoto, T. Miyawaki, Y. Matsushita, T. Okada, H. Natsugari, and M. Kida, J. Antibiotics, 34, 171 (1981); R. Reiner, U. Weiss, U. Brombacher, P. Lanz, M. Montavon, A. Furlenmeier, P. Angehrn, and P. J. Probst, J. Antibiotics, 33, 783 (1980).

25. C. H. O'Callaghan, D. G. H. Livermore, and C. E. Newall,
    German Offen. DE 2,921,316 (1979); Chem. Abstr., 92,
    198413c (1980).

26. T. Takaya, H. Takasugi, K. Tsuji, and T. Chiba, German
    Offen. DE 2,810,922 (1978); Chem. Abstr., 90, 204116k
    (1979).

27. H. Otsuka, W. Nagata, M. Yoshioka, M. Narisada, T.
    Yoshida, Y. Harada, and H. Yamada, Med. Res. Revs., 1,
    217 (1981); M. Narisada, H. Onoue, and W. Nagata,
    Heterocycles, 7, 839 (1977); M. Narisada, T. Yoshida, O.
    Onoue, M. Ohtani, T. Okada, T. Tsugi, I. Kikkawa, N. Haga,
    H. Satoh, H. Itani, and W. Nagata, J. Med. Chem., 22, 757
    (1979).

13 Miscellaneous Fused Heterocycles

Medicinal agents discussed to this point have been roughly classifiable into some common structural groups; biological activity often followed the same rough classification. As was the case in the preceding volumes in this series, a sizable number of compounds, often based on interesting heterocyclic systems, defy ready grouping by structure. These are thus discussed below under the cover of "miscellaneous." It might be added as an aside that this section may include compounds that will someday move to new chapters. If one of these drugs proves to be a major clinical or marketing success, it will no doubt occasion a considerable amount of competitive work. Since some of this work will undoubtedly result in agents with generic names, the class may well finally grow to the point where it will require listing as a structural group.

A rather simple derivative of imidazoimidazoline has been described as an antidepressant agent. Preparation of this compound starts with the displacement of the nitramine grouping in imidazoline derivative 1 by phenylethanolamine 2. The product of this reaction is then treated with thionyl chloride. The probable chloro intermediate (4) cyclizes under the reaction conditions to afford imafen (5).[1]

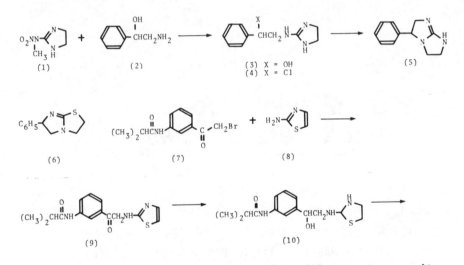

The imidazothiazoline tetramisole (6) has shown quite good activity as a broad spectrum anthelmintic agent. This drug has in addition aroused considerable interest as an agent which modifies the host immune response. Further substitution on the aromatic ring has proved compatible with activity. Displacement of halogen on the phenacyl bromide 7 with aminothiazole 8 affords the alkylated product 9. Catalytic hydrogenation serves to reduce both the heterocyclic ring and the carbonyl group (10). Cyclization by means of sulfuric acid completes the synthesis of butamisole (11).[2]

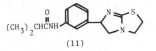

(11)

Benzofurans of the very general structure represented by 12 have formed the basis of several quite effective drugs for treatment of cardiovascular disease. It is thus of note that replacement of the aromatic nucleus by the isosteric indolizidine system affords a compound with quite similar activity. Friedel-Crafts type acylation of indolizidine 13 with substituted benzoyl chloride 14 gives the ketone (15). Removal of the protecting group gives the free phenol. Alkylation by means of N,N-di(n-butyl)-2-chloro-ethylamine affords the corresponding basic ether. There is thus obtained the antiarrhythmic agent butoprizine (17).[3]

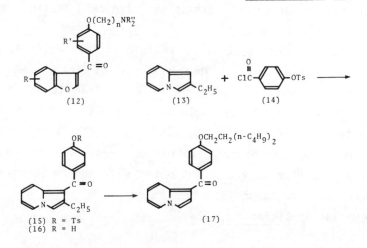

(12)   (13)   (14)

(15) R = Ts
(16) R = H   (17)

Aggregation of blood platelets is the requisite first event for the maintenance of intact circulation in the face of any break in a blood vessel. It is the platelet clump that starts the long and complicated process leading to closure of the broken vessel by an organized blood clot. Though this property of platelets is vital to maintenance of the circulatory system, an excessive tendency to aggregation can also lead to problems. Thus platelet clumps formed in blood vessels in the absence of injury can lead to blockade of blood circulation and subsequent injury. Strokes and some types of myocardial infarcts have thus been associated with platelet clumps. The nonsteroid antiinflammatory agents as a class show platelet antiaggregation activity in a number of test systems; however, there has been a considerable amount of effort expended to uncovering agents from other structural classes that will not share the deficits of the nonsteroid antiinflammatories. Ticlopidine (24), a drug that shows good activity in various animal models has undergone extensive clinical testing as a platelet antiaggregator.

The key intermediate 21 is in principle accessible in any of several ways. Thus reaction of thiophenecarboxaldehyde 18 with amninoacetal 19 would lead to the Schiff base 20; treatment with acid would result in formation of the fused thiophene-pyridine ring (21). Alkylation of that intermediate with benzyl chloride 22 gives the corresponding ternary iminium salt 23. Treatment with sodium borohydride leads to reduction of the quinolinium ring and thus formation of ticlopidine (24).[4]

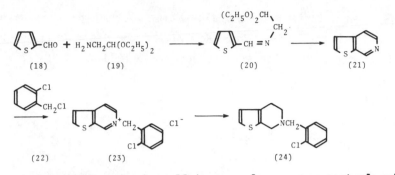

The purines, as is well known, play a very central role in the biochemistry of life. This heterocyclic nucleus is involved in vital processes in a host of guises, from its participation in the genetic message to its part in the energy transmission system and perhaps even as a neuro-transmitter. It is thus not surprising that considerable attention has been devoted to this heterocyclic system as a source for drugs; it is somewhat unexpected that so few of these efforts have met with success.

The success of antibacterial therapy hinges largely on the fact that the metabolism of bacteria differs sufficient-ly from that of the host so that it is possible to interfere selectively with this process. Viral infections have been much more difficult to treat because the organism in effect takes over the metabolic processes of the host cell; selectivity is thus very slight. One of the signal breakthroughs in this field of therapy is an agent that takes advantage of one of those small differences, acting as a false substrate for a biochemical process necessary for viral replication. It is pertinent that this drug, acyclovir (27) may be viewed as an analogue of the nucleoside guanosine 28, in which two of the ring carbons of ribose (or deoxyribose) have been deleted. Preparation of

this agent starts with the alkylation of guanine (25) with
the chloromethyl ether 25a. Removal of the protecting group
(26) by saponification affords acyclovir (27).[5]

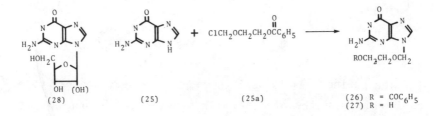

The uric acid derivative theophylline (29) is one of
the mainstays as a bronchodilator drug for the treatment of
asthma. This agent's narrow therapeutic index and host of
side effects has led to an active search for a safer
derivative. Synthesis of one such compound starts with the
condensation of amine 30 with methyl isocyanate. Acylation
of the resulting urea (31) with cyanoacetic acid gives the
intermediate 32; this is then cyclized to the corresponding
uracil 33 by means of base. Nitrosation (34) followed by
reduction of the newly introduced nitroso group gives the
ortho diamine function (35). The remaining ring is
constructed by first acylating 35 with acetic anhydride (to
give 36); cyclization again by means of base completes the
purine nucleus. There is thus obtained the bronchodilating
agent verofylline (37).[6]

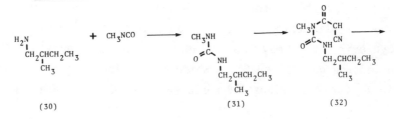

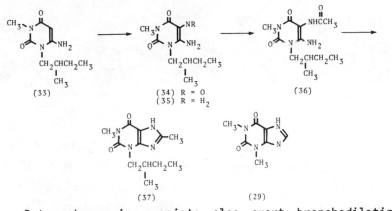

Beta adrenergic agonists also exert bronchodilating effects. These drugs are thus often used in conjunction with theophiline in asthma therapy. A drug that combines both moieties, reproterol (40), has interestingly proved clinically useful as an antiasthmatic agent. This compound can in principle be obtained by first alkylating theophylline with 1-bromo-3-chloropropane to give 38. Use of this halide to alkylate aminoalcohol 39 would then afford reproterol (40).

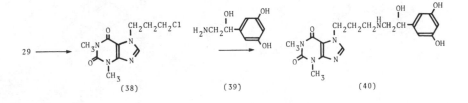

As noted earlier (see Chapter 10), 4-acylpiperidines separated from benzimidazole by a three carbon chain often show antipsychotic activity. The heterocycle can apparently be replaced by a pyridopyrimidine ring. Thus alkylation of piperidine 41 with halide 42 affords pirenperone (43).[7]

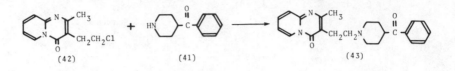

Hydrazinopyridazines such as hydralazine have a venerable history as antihypertensive agents. It is of note that this biological activity is maintained in the face of major modifications in the heterocyclic nucleus. The key intermediate keto ester 45 in principle can be obtained by alkylation of the anion of piperidone 44 with ethyl bromoacetate. The cyclic acylhydrazone formed on reaction with hydrazine (46) is then oxidized to give the aromatized compound 47. The hydroxyl group is then transformed to chloro by treatment with phosphorus oxychloride (48). Displacement of halogen with hydrazine leads to the formation of endralazine (49).[8]

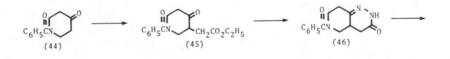

Two closely related pyridotriazines have been described as antifungal agents. Displacement of halogen on nitrochloropyridine 50 with the monocarbamate of hydrazine affords intermediate 51. This is then first hydrolyzed to the free hydrazine (52) and the nitro group reduced to the corresponding amine (53). Condensation of this intermediate with phenylacetic acid leads to formation of the cyclic amidine derivative 54. Oxidation with manganese dioxide introduces the remaining unsaturation; there is thus obtained triafungin (55).[9] Condensation of 53 with phenoxyacetic acid gives, after aromatization of the first formed product, the antifungal agent oxyfungin (56).[10]

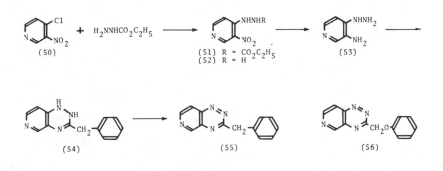

The enormous commercial success of the benzodiazepine anxiolytic agents has spurred a correspondingly large effort in many laboratories aimed at developing novel analogues (see, for example, Chapter 11). In this case it is probably no exaggeration to say that every part of the parent molecule has been modified in the search for novel patentable analogues. In the course of such work it has been found that replacement of the fused benzene ring by a heterocyclic ring is compatible with tranquilizing activity.

Preparation of one of the analogues in which benzene is replaced by pyrazole starts by nitration of pyrazole carboxylic acid 57. The product, 58, is then converted to the acid chloride (59). This intermediate is then used to acylate benzene in a Friedel-Crafts reaction. The nitroketone is then reduced to the corresponding amine. Reaction with ethyl glycine can be visualized as involving initially formation of the Schiffs' base (62). Displacement of ethoxide by the ring amino group leads to formation of the lactam. There is thus obtained ripazepam (63).[11]

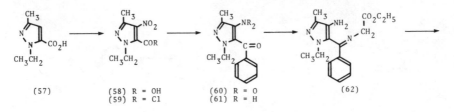

    (57)   (58) R = OH   (60) R = O   (62)
                   (59) R = Cl   (61) R = H

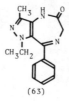

(63)

A somewhat different strategy is employed for preparation of the desoxy analogue containing the reversed pyrazole. Acylation of chloropyrazole 64 with m-chloro-benzoyl chloride affords the ketone 65. Reaction of that with ethylenediamine leads directly to the anxiolytic agent zometapine (66).[12] The overall sequence obviously involves sequential Schiff base formation and nucleophilic displacement of chlorine; the order of these steps is not clear.

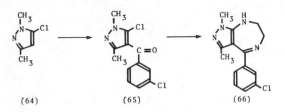

(64)   (65)   (66)

In a similar vein, acylation of aminoketone 67 with chloroacetyl chloride affords the corresponding chloroamide 68. Reaction of that intermediate with ammonia serves to form the diazepine ring, possibly via the glycinamide. The product bentazepam (69) is described as a tranquilizer.[13]

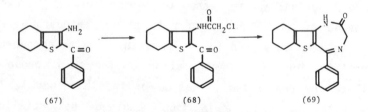

(67)   (68)   (69)

Carboxylic acid derivatives of heterocycles have proved a source of compounds that show the same allergic mediator release inhibiting activity as sodium cromoglycate. A number of these agents have been taken to the clinic for trial as antiallergic agents.

Friedel-Crafts cyclization of phenoxy ether 70 leads to the corresponding xanthone 71. Exhaustive oxidation of the methyl group leads to the carboxyllic acid, xanoxate (72).[14]

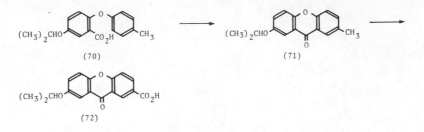

Preparation of the analogue in which isopropyloxy is replaced by a methylsulfoxide involves a somewhat more complex scheme. Aromatic nucleophilic displacement of halogen in dicarboxyl ester <u>73</u> leads to diphenyl ether <u>75</u>. The product is then saponified (<u>76</u>), cyclized to the xanthone and again esterified (<u>78</u>). The aromatic ether is then demethylated to the free phenol (<u>79</u>). This group is converted to the thiocarbamate <u>80</u> by means of dimethylthiocarbamoyl chloride. Thermal rearrangement of the thiocarbamate function by the method of Newman results in overall exchange of sulfur for oxygen to afford thiocarbamate <u>81</u>. This is then converted to the free thiol, with accompanying saponification (<u>82</u>). Methylation of the thiol group (<u>83</u>) followed by controlled oxidation of the thioether leads to the sulfoxide. There is thus obtained the antiallergic agent <u>tixanox</u> (<u>84</u>).[15]

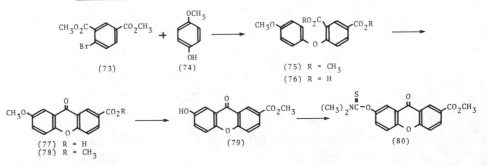

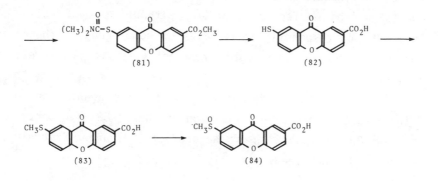

(81)   (82)

(83)   (84)

It has by now been well established that the tricyclic
ring system of the phenothiazine tranquilizers is not an
absolute requirement for antipsychotic activity; that moiety
has been successfully replaced by ring systems as diverse as
acridine and even dihydroanthracene. It should thus not be
surprising to note that dibenzopyran derivatives also lead
to active compounds. Thus reaction of xanthone 85 with the
Grignard reagent from chloropiperidine 86 gives after dehydration the antipsychotic agent clopipazam (87).[16]

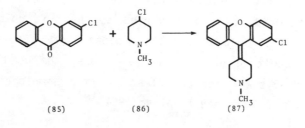

(85)   (86)   (87)

It is by now apparent that the nature of the aryl group
in the arylacetic and arylpropionic acid antiinflammatory
agents can be varied quite widely without loss of activity. The corresponding derivatives of homologous xanthones
and thioxanthones thus both show activity as nonsteroid
antiinflammatory agents.

Starting material for the first of these agents can in
principle be obtained by alkylation of phenol 88 with benzyl
chloride 89. Cyclization of the product (90) under Friedel-
Crafts conditions leads directly to isoxepac (91).[17]

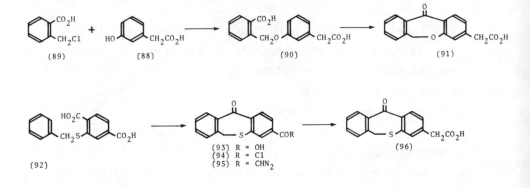

Preparation of the sulfur analogue involves as the
first step cyclization of the terephthalic acid derivative
92. The acid is then converted to the acid chloride and
this is allowed to react with diazomethane. Rearrangement
of the resulting diazoketone (95) under the conditions of
the Arndt-Eistert reaction leads to the homologated acid.
There is thus obtained tiopinac (96).[18]

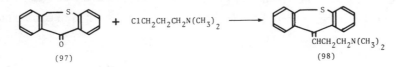

(97)   (98)

Antidepressant agents show almost the same degree of tolerance as to the nature of the tricyclic moiety as do the antipsychotic agents. Thus the dehydration product(s) from the condensation of ketone 97 with the Grignard reagent from 3-chloroethyl-N,N-dimethylamine affords the antidepressant diothiepin (98).[19]

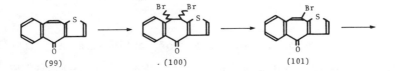

(99)   . (100)   (101)

Historically, both the tricyclic antipsychotic and antidepressant agents are derived in almost direct line from a series of tricyclic antihistaminic compounds (see 104 below). Minor changes in structure in some of the newer compounds in fact lead to drugs in which antihistaminic activity predominates. Thus ketotifen, which differs from antipsychotic compounds such as 87 only in detail, is a rather potent antihistamine. Bromination of ketone 99 occurs on the ethylene bridge to afford the 1,2 dibromide as a mixture of isomers (100); dehydrohalogenation by means of strong base gives the vinyl bromide 101 apparently as a single regioisomer. Reaction with the Grignard reagent from N-methyl-4-bromopiperidine gives the alcohol 102. Exposure

of the intermediate to strong acid leads to dehydration of
the alcohol and hydrolysis of the vinyl bromide to the
corresponding ketone.   There is thus obtained ketotifen
(103).[20]

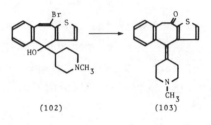

(102)   (103)

Since many of the uses of antihistamines involve
conditions such as rashes, which should be treatable by
local application, there is some rationale for developing
drugs for topical use.   The known side effects of antihistamines could in principle be avoided if the drug were
functionalized so as to avoid systemic absorption.   The
known poor absorption of quaternary salts make such derivatives attractive for nonabsorbable antihistamines for
topical use.   Thus, reaction of the well-known antihistaminic drug promethazine (104) with methyl chloride leads
to thiazinium chloride (105).

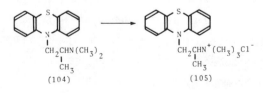

(104)   (105)

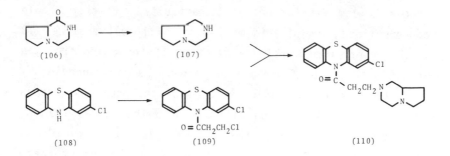

(106)　(107)　(110)

(108)　(109)

Attachment of the basic side chain to the phenothiazine nucleus by means of a carbonyl function apparently abolishes the usual CNS or antihistamine effects shown by most compounds in this class. The product <u>azaclorzine</u> instead is described as an antianginal agent. Reduction of proline derivative <u>106</u> with lithium aluminum hydride gives the corresponding fused piperazine <u>107</u>. Use of that base to alkylate the chloroamide <u>109</u>, obtained from acylation of phenothiazine with 3-chloropropionyl chloride, leads to <u>azaclorzine</u> (<u>110</u>).[21]

(112)　(111)　(113)

Fluorobutyrophenone derivatives of 4-arylpiperidines are well-known antipsychotic agents. It is thus interesting to note that the piperidine can in fact be fused onto an indole moiety with retention of activity. Fischer indole condensation of 4-piperidone <u>111</u> with phenylhydrazine <u>112</u> leads to the indole <u>113</u>. Alkylation of the anion from the indole with p-bromofluorobenzene gives the corresponding N-arylated derivative (<u>114</u>). Removal of the protecting group followed by alkylation on nitrogen with the acetal from p-p-fluorobutyryl chloride gives intermediate <u>116</u>. Hydrolysis of the acetal followed by reduction of the ketone by means of sodium borohydride gives the antipsychotic agent <u>flutroline</u> (<u>118</u>).[22]

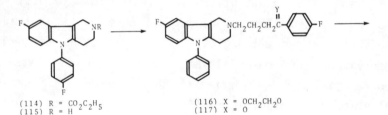

(114) R = CO$_2$C$_2$H$_5$
(115) R = H

(116) X = OCH$_2$CH$_2$O
(117) X = O

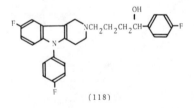

(118)

A remarkably simple fused indole devoid of the traditional side chains is described as an antidepressant agent. Michael addition of the anion from indole ester <u>119</u> to acrylonitrile affords the cyanide <u>120</u>. Selective reduction of the nitrile leads to the aminoester <u>121</u>. This is then cyclized to the lactam (<u>122</u>). Reduction of the carbonyl group by means of lithium aluminum hydride leads to azepindole (<u>123</u>).[23]

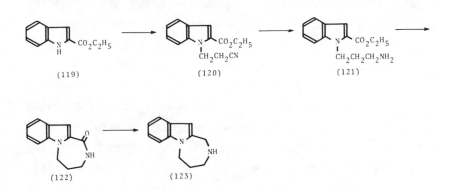

(119)   (120)   (121)

(122)   (123)

A fused pyrazoloquinolone provides an exception to the
rule that antiallergic agents must contain a strongly acidic
proton. Entry to the ring system is gained by electrocyclic
reaction of diazoindolone 124 (possibly obtained by reaction
of the anion from indolone with p-toluenesulfonyl azide)
with propargylaldehyde. The initial adduct to the 1,3-
dipole represented by the diazo group can be formulated as
the spiro intermediate 125. Bond reorganization would then
lead to the observed product (126). Reduction of the
carbonyl with sodium borohydride leads to the corresponding
alcohol, and thus pirquinozol (127).[24]

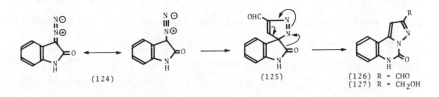

(124)   (125)   (126) R = CHO
                                                  (127) R = CH$_2$OH

An imidazoquinazoline constitutes still another compound that does not fall in the classification of a nonsteroid antiinflammatory agent yet shows good platelet antiaggregating activity. Condensation of benzyl chloride 128 with the ethyl ester of glycine gives alkylated product 129. Reduction of the nitro group leads to aniline 130. Reaction with cyanogen bromide possibly gives cyanamide 131 as the initial intermediate. Addition of aliphatic nitrogen would then lead to formation of the quinazoline ring (132). Amide formation between the newly formed imide and the ester would then serve to form the imidazolone ring. Whatever the details of the sequence, there is obtained in one step anagrelide (133).[25]

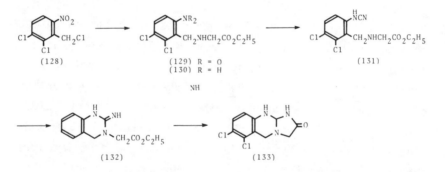

A seemingly complex heterocycle which on close exam-ination is in fact a latentiated derivative of a salicylic acid shows antiinflammatory activity. It might be spec-ulated that this compound could quite easily undergo meta-bolic transformation to a salicylate and that this product is in fact the active drug. Condensation of acid 134 with hydroxylamine leads to the hydroxamic acid 135. Reaction of that with the ethyl acetal from 4-chlorobutyraldehyde then leads to the cyclic carbinolamine derivative 136. Treatment with mild base causes internal alkylation and consequent formation of the last ring.   There is thus obtained meseclazone (137).[26]

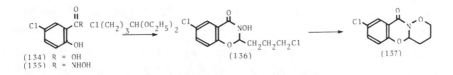

(134) R = OH
(135) R = NHOH

(136)

(137)

The observation that a carboxyl derivative of a pyrimidinoquinoline shows mediator release inhibiting activity is in consonance with the earlier generalization. Knoevenagel condensation of nitroaldehyde 138 with cyano-acetamide gives the product 139.   Treatment with iron in acetic acid leads to initial reduction of the nitro group (140).   Addition of that function to the nitrile leads to formation of the quinoline ring (141).   Reaction of that compound with ethyl oxalate results in formation of the quinazoline ring.   The product, pirolate (142), is described as an antiallergy agent.[27]

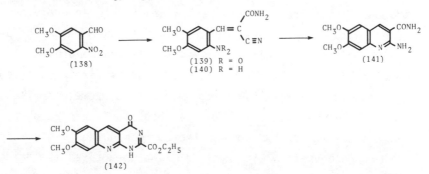

(138)

(139) R = O
(140) R = H

(141)

(142)

A tetracyclic heterocycle that bears little relation to
any clinically used drug has been described as an antiinflammatory agent.   The compound is prepared in rather
straightforward manner by initial condensation of dihalide
143 with 1,2-diaetcylhydrazine.   Hydrolysis of this product
gives cyclic hydrazone 145.   Exposure to a second mole of
dihalide leads to diftalone (146).[28] The regiochemistry may
be rationalized by assuming that the more reactive acid
chloride attacks the more nucleophilic unacylated nitrogen.

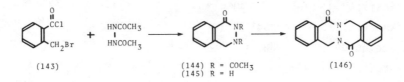

(143)   (144) R = COCH$_3$ (146)
                               (145) R = H

The tricyclic antidepressants (as well as, incidentally
the antipsychotic drugs) are characterized by a three carbon
chain between the ring system and the basic nitrogen.   Incorporation of one of those carbon atoms into an additional
fused ring is apparently consistent with activity.   Preparation of this compound involves first homologation of the
side chain.   Thus the carboxylic acid 147 is first converted
to the acid chloride (148); reaction with diazomethane leads
to the   diazoketone 149.   This is then   subjected to photo-
lytic rearrangement to afford the corresponding acetic acid
(150).   Condensation with methylaniline then gives the amide
151.   Reduction with lithium aluminum hydride affords

azipramine (152).[29]

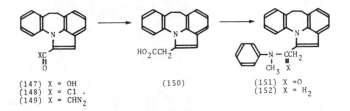

(147) X = OH
(148) X = Cl .
(149) X = CHN$_2$

(150)

(151) X =O
(152) X = H$_2$

Any migraine sufferer will willingly testify that this
condition has little in common with the headaches to which
the rest of mankind are subject. Recent medical studies too
have shown fairly conclusively that, whatever the etiology
of migraine, it is a condition quite distinct from the
common headache. The syndrome is in fact so distinct as to
be untouchable by the common headache cures such as
aspirin. Drugs for treatment of migraine are unfortunately
almost nonexistent. (The lack of appropriate animal models
in no small way hinders the search for a treatment.)   A
benzofuranobenzoxepin has interestingly been described as an
antimigraine agent. Bromination of benzofuran 153 proceeds
on the methyl group to give the arylmethyl bromide 154.
Displacement by phenoxide then leads to intermediate 155.
Saponification (156) followed by Friedel-Crafts cyclization
serves to form the seven-membered ring (157). Condensation
of the ketone with the Grignard reagent from 3-chloropropyl-
N,N-dimethylamine gives the olefin on dehydration, possibly
as a mixture of isomers. There is thus obtained oxetorene
(158).[30]

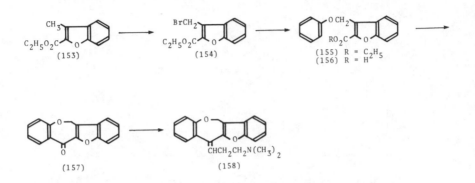

In the steroid series, hormone antagonists usually bear some structural resemblance to the endogenous agonists. That is to say, antagonists are almost always steroids themselves. Even in the case of the nonsteroid estrogen antagonists, there is a fairly clear structural resemblance to estradiol. It is thus somewhat surprising to note a clearly nonsteroidal androgen antagonist. The compound in question, pentomone (163) is, as a result of this activity, a potential drug for treatment of prostate enlargement. Condensation of salicylaldehyde 159 with cyclohexanone 160 proceeds twice to give directly the pentacyclic intermediate 161. The reaction may be visualized as initial conjugate addition of phenoxide to the enone followed by interception of the resulting anion by the aldehyde carbonyl group. Hydrogenation of the intermediate reduces both the double bonds and the carbonyl group (162). Back oxidation of the alcohol thus formed with pyridinium chlorochromate affords pentomone (163).

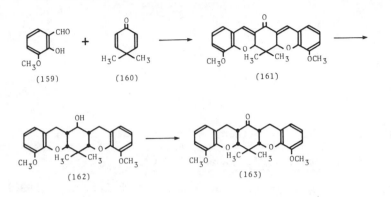

(159)   (160)   (161)

(162)   (163)

The ergolines have provided a number of drugs that show interaction with neurotransmitters. Depending on the substitution pattern, they may be dopamine agonists or antagonists, α-adrenergic blockers, or inhibitors of the release of prolactin. A recent member of the series, <u>pergolide</u> (<u>167</u>), shows activity as a dopamine antagonist. Reduction of ester <u>164</u>[31] by means of lithium aluminum hydride gives the corresponding alcohol; this is then converted to its mesylate (<u>166</u>). Displacement with methanethiol affords pergolide (<u>167</u>).[32]

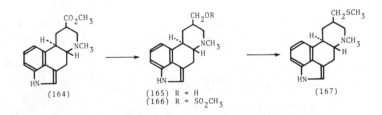

(164)   (165) R = H   (167)
                     (166) R = SO$_2$CH$_3$

REFERENCES

1. J. L. H. Van Gelder, A. H. M. Raeymaekers, L. F. C. Roevens and W. J. Van Laerhoven, U.S. Patent 3,925,383; Chem. Abstr., 84, 180264e (1976).

2. L. D. Spicer and J. J. Hand, French Demande 2,199,979; Chem. Abstr., 82, 16835e (1975).

3. J. Gubin and G. Rosseels, German Offen 2,707,048; Chem. Abstr., 88, 6719e (1978).

4. J. P. Maffrand and F. Elloy, Eur. J. Med. Chem., 9, 483 (1974).

5. H. J. Shaeffer, German Offen., 2,539,963 (1976).

6. J. Diamond, German Offen., 2,713,389; Chem. Abstr., 88, 22984t (1978).

7. L. E. J. Kennis and J. C. Mertens, U.S. Patent 4,347,287; Chem. Abstr., 98, 16716 (1983).

8. E. Schenker, Swiss Patent 565,797; Chem. Abstr., 83, 206311z (1975).

9. G. C. Wright, A. V. Bayless, and J. E. Gray, German Offen., 2,427,382; Chem. Abstr., 82, 171087f (1975).

10. G. C. Wright, A. V. Bayless, and J. E. Gray, German Offen., 2,427,377; Chem. Abstr., 82, 171088g (1975).

11. I. C. Nordin, U.S. Patent 3,553,207; Chem. Abstr., 75, 5972b (1971).

12. H. A. DeWald and S. J. Lobbestael, South African Patent 73 07696; Chem. Abstr., 84, 59594j (1976).

13. F. J. Tinney, U.S. Patent 3,558,606; Chem. Abstr., 74, 141896m (1971).

14. D. E. Bays, German Offen., 2058295; Chem. Abstr., 75, 98447x (1971).

15. J. R. Pfister, I. T. Harrison, and J. H. Fried, Chem. Abstr., 85, 21108g (1976).

16. C. L. Zirkle, German Offen., 2,549,841; Chem. Abstr., 85, 78025m (1976).

17. P. Herbst and D. Hoffmann, German Offen., 2,600,768; Chem. Abstr., 85, 143001s (1976).

18. J. Ackrell, Y. Antonio, F. Fidenico, R. Landeros, A. Leon, J. M. Muchowski, M. L. Maddox, P. H. Nelson, and W. H. Rooks, J. Med. Chem. 21, 1035 (1978).

19. C. L. Zirkle, U.S. Patent 3,609,167; Chem. Abstr., 75, 151694d (1971).

20. J. P. Bourquin, G. Schwarb, and E. Waldvogel, German Offen. 2,144,490; Chem. Abstr., 77, 34296f (1972).

21. N. V. Kaverina, G. A. Markova, G. G. Chichkanov, L. S. Nazarovo, A. M. Likhosherstor, and A. P. Skoldinov, Khim. Pharm. Zh., 12, 97 (1978)

22. J. J. Plattner, C. A. Harbert, J. R. Tretter, and W. M. Welch Jr., German Offen., 2,514,084; Chem. Abstr., 84, 44008x (1976).

23. B. Reynolds and J. Carson, German Offen. 1928726; Chem. Abstr., 72, 55528v (1970).

24. B. R. Vogt, German Offen. 2,726,389; Chem. Abstr., 88, 121240d (1978).

25. W. N. Beverung and R. A. Partyka, U.S. Patent 3,932,407; Chem. Abstr., 84, 10564 (1976).

26. D. B. Reisner, B. J. Ludwig, H. M. Bates, and F. M. Berger, German Offen., 2010418; Chem. Abstr., 73, 120644s (1970).

27. T. H. Althuis, L. J. Czuba, H. J. E. Hess, and S. B. Kadin, German Offen., 2,418,498; Chem. Abstr., 82, 73015m (1975).

28. E. Bellasio and E. Testa, Il Farmaco, Ed. Sci., 25, 305 (1970).

29. M. Riva, L. Toscano, A. Bianchetti, and G. Grisanti, German Offen., 2,529,792; Chem. Abstr., 84, 150530w (1976).

30. F. Binon and M. Descamps, German Offen. 1963205; Chem. Abstr., 73, 77221n (1970).

31. D. Lednicer and L. A. Mitscher, "The Organic Chemistry of Drug Synthesis", Vol. 2, Wiley, New York, 1980, p. 479.

32. E. C. Kornfeld and N. J. Bach, Eur. Patent Appl., 3,667; Chem. Abstr., 92, 181450q (1980).

Cross Index of Drugs

Antiacne

Etretinate   Montretinide
Isotretinoin   Piroctone

Antiallergic

Isamoxole   Pirolate
Lodoxamide ethyl   Pirquinozol
Nivimedone   Tixanox
Oxatomide   Xanoxate

Antiamebic

Quinfamide

Antianginal

Azaclorzine   Molsidomine
Bepridil   Nicarpidine
Cinepazet   Nicordanil
Diltiazem   Nimodipine
Droprenilamine   Tosifen

Antiarrhythmic

Butoprozine   Lorcainide
Clofilium Phosphate   Meobentine
Disobutamine   Oxiramide
Drobuline   Pirmenol
Encainide   ·   Ropitoin
Emilium Tosylate   Tocainide
Flecainide

Antibacterial

Droxacin   Metioprim
Fludalanine   Rosoxacin
Flumequine   Tetroxoprim

Antibiotic

Amidinocillin   Cefroxadine
Amidinocillin Pivoxyl   Cefsulodin
Azolocillin   Ceftazidine

Bacampicillin
Cefaclor
Cefaparole
Cefatrizine
Cefazaflur
Cefoperazone
Cefonicid
Ceforanide
Cefotaxime
Cefotiam

Ceftizoxime
Cefuroxime
Mezlocillin
Moxlactam
Piperacillin
Pirbencillin
Piridicillin
Sarmoxicillin
Sarpicillin

Anticonvulsant

Cinromide

Antidepressant

Azepindole
Azipramine
Cyclopenzaprine
Cyclindole
Dothiepin
Fluotracen
Fluoxetine
Imafen

Napactidine
Nisoxetine
Nitrafudam
Pridefine
Tametraline
Viloxazine
Zimelidine

Antidiarheal

Nufenoxole

Antiemetic

Domperidone   Nabilone

Antifungal

Azoconazole
Butoconazole
Doconazole
Ketoconazole
Naftidine
Orconazole
Oxifungin

Parconazole
Sulconazole
Terconazole
Tioconazole
Tolciclate
Triafungin

Antihelmintic

Butamisole   Frentizole

Carbantel   Nocodazole
Felsantel   Tioxidazole
Fenbendazole

Antihistamine

Astemizole   Thiazinium Chloride
Ketotifen

Antihypertensive

Captopril   Proroxan
Endralazine   Terazocin
Guanfacine   Tiamenidine
Indorenate   Tiodazocin
Ketanserin

Antihypertensive - β-Blocker

Bucindolol   Penbutolol
Diacetolol   Primidolol
Exaprolol   Prizidolol
Pamatolol

Antihypertensive - α,β-Blocker

Bevantolol   Medroxalol
Labetolol   Sulfinalol

Anti-inflamatory - Steroid

Acolmethasone Dipropionate   Haloprednone
Budesonide   Meclorisone Dibutyrate
Ciprocinonide   Procinonide
Flumoxonide

Anti-inflamatory - Non-Steroid

Anitrazafen   Indoprofen
Amefenac   Isoxepace
Bromperamole   Oxarbazole
Carprofen   Meseclazone
Diftalone   Morniflumate
Epirizole   Orpanoxin
Fencolofenac   Pirazolac
Fenclosal   Sermetacin
Floctafenine   Talniflumate

Fluquazone
Fluproquazone
Fluretofen

Tiopinac
Zidometacin
Zomepirac

Antimalarial

Halofantrine

Antimigraine

Oxetorone

Antineoplastic

Ametantrone
Cyclophosphamide
Estramustine
Etoprine
Ifosfamide

Metoprine
Mitoxantrone
Prednimustine
Tegafur
Trofosfamide

Antiprotozoal

Bamnidazole
Ornidazole

Misonidazole

Antipsychotic

Clopipazam
Cloroperone
Declenperone
Flucindole

Fluotracen
Flutroline
Halopemide
Pipenperone

Antiulcer

Arbaprostil
Etintidine
Oxmetidine

Ranitidine
Tolimidone

Antiviral

Acyclovir
Arildone

Enviroxime

Anxiolytic

Adinazolam
Bentazepam

Lormetazepam
Midazolam

Brofoxine   Quazepam
Caroxazone   Ripazepam
Elfazepam   Tioperidone
Fosazepam   Zometapine

Bronchodilator

Ipatropium Bromide   Verofylline

Brochoclilator - β-Adrenergic

Bitolterol   Nisobuterol
Colterol   Prenalterol
Carteolol   Reproterol
Dipirefrin

Cardiotanic

Actodigin   Butopamine
Amrinone   Carbazeran

Catract Inhibitor

(Aldose Reductase Inhibitors)

Alrestatin   Sorbinil

Cognition Enhancer

Amacetam

Diagnostic Aid (Pancreatic Function)

Bentiromide

Diuretic

Azosemide   Muzolimine
Fenquizone   Ozolinone
Indacrinone   Piretanide

Dental Carries Prohylactic

Ipexidine

Dopamine Antagonist

Pergolide

Estrogen Antagonist

Nitromifene   Trioxifene
Tamoxifen

Hypoglycemic

Glicetanile   Pirigliride
Gliflumide

Hypolipidemic

Benzafibrate   Gemcadiol
Cetaben   Gemfibrozil
Ciprofibrate

Immunomodulator

Azarole

Muscle Relaxant

Clodanolene   Xilobam
Lidamidine

Uterine Stimulant/Oxytocic

Carboprost   Sulprostone
Mefenprost

Peripheral Vasodilator

Buterizine   Suloctidil
Cetiedil   Tipropidil

Platelet Agregation Inhibitor

Anagrelide   Ticlopidine
Epoprostenol

Progestin

Gestodene   Gestrinone

Sedative

Fenobam   Milenperone

Vasosilator

Alprostadil   Epoprostenol

Vitamin (D$_3$)

Calcifediol   Calcitriol

Cumulative Index, Vols. 1-3

Index